ENVIRONMENTAL SCIENCE, ENGINEERING AND TECHNOLOGY

CARBON CONSIDERATIONS

BIOCHAR, BIOMASS, BIOPOWER, AND SEQUESTRATION

Environmental Science, Engineering and Technology

Additional books in this series can be found on Nova's website under the Series tab.

Additional E-books in this series can be found on Nova's website under the E-book tab.

Climate Change and its Causes, Effects and Prediction

Additional books in this series can be found on Nova's website under the Series tab.

Additional E-books in this series can be found on Nova's website under the E-book tab.

ENVIRONMENTAL SCIENCE, ENGINEERING AND TECHNOLOGY

CARBON CONSIDERATIONS

BIOCHAR, BIOMASS, BIOPOWER, AND SEQUESTRATION

WESLEY A. GREEN
AND
LORI G. WAYMAN
EDITORS

nova publishers
New York

For permission to use material from this book please contact us:
Telephone 631-231-7269; Fax 631-231-8175
Web Site: http://www.novapublishers.com

Library of Congress Cataloging-in-Publication Data

ISBN: 978-1-62618-435-0

Published by Nova Science Publishers, Inc. † New York

CONTENTS

PREFACE

Widespread concern about global climate change has led to interest in reducing emissions of carbon dioxide (CO_2) and, under certain circumstances, in counting additional carbon absorbed in soils and vegetation as part of the emissions reductions. Congress may consider options to increase the carbon stored (sequestered) in forests as it debates this and related issues. Forest are a significant part of the global carbon cycle. Plants use sunlight to convert CO_2, water, and nutrients into sugars and carbohydrates, which accumulate in leaves, twigs, stems, and roots. Plants also respire, releasing CO_2. Plants eventually die, releasing their stored carbon to the atmosphere quickly or to the soil where it decomposes slowly and increases soil carbon levels. Little information exists on the processes and diverse rates of soil carbon change. This book examines basic questions concerning carbon sequestration with a focus on biochar for soil fertility and natural carbon sequestration in forests; and biomass feedstocks for biopower.

Chapter 1 - Widespread concern about global climate change has led to interest in reducing emissions of carbon dioxide (CO_2) and, under certain circumstances, in counting additional carbon absorbed in soils and vegetation as part of the emissions reductions. Congress may consider options to increase the carbon stored (sequestered) in forests as it debates this and related issues.

Forests are a significant part of the global carbon cycle. Plants use sunlight to convert CO_2, water, and nutrients into sugars and carbohydrates, which accumulate in leaves, twigs, stems, and roots. Plants also respire, releasing CO_2. Plants eventually die, releasing their stored carbon to the atmosphere quickly or to the soil where it decomposes slowly and increases soil carbon levels. However, little information exists on the processes and diverse rates of soil carbon change.

How to account for changes in forest carbon has been contentious. Land use changes—especially afforestation and deforestation—can have major impacts on carbon storage. Foresters often cut some vegetation to enhance growth of desired trees. Enhanced growth stores more carbon, but the cut vegetation releases CO_2; the net effect depends on many factors, such as prior and subsequent growth rates and the quantity and disposal of cut vegetation. Rising atmospheric CO_2 may stimulate tree growth, but limited availability of other nutrients may constrain that growth.

In this context, timber harvesting is an especially controversial forestry practice. Some argue that the carbon released by cutting exceeds the carbon stored in wood products and in tree growth by new forests. Others counter that old-growth forests store little or no additional carbon, and that new forest growth and efficient wood use can increase net carbon storage. The impacts vary widely, and depend on many factors, including soil impacts, treatment of residual forest biomass, proportion of carbon removed from the site, and duration and disposal of the products. To date, the quantitative relationships between these factors and net carbon storage have not been established.

Some observers are concerned that "leakage" will undermine any U.S. efforts to sequester carbon by protecting domestic forests. By leakage, they mean that wood supply might shift to other sites, including other countries, exacerbating global climate change and causing other environmental problems, or that wood products might be replaced by other products that use more energy to manufacture (thus releasing more CO_2). Others counter that the "leakage" arguments ignore the enormous disparity in ecological systems and product preferences, and discount possible technological solutions.

Several federal government programs affect forestry practices and thus carbon sequestration. Activities in federal forests affect carbon storage and release; timber harvesting is the most controversial such activity. Federal programs also provide technical and financial help for managing and protecting private forests, and tax provisions affect private forest management. Various federal programs can also affect the extent of forested area, by supporting development (which may cause deforestation) or encouraging tree planting in open areas, such as pastures.

Global climate change is a widespread and growing concern that has led to extensive international discussions and negotiations. Responses to this concern have focused on reducing emissions of greenhouse gases, especially carbon dioxide, and on measuring carbon absorbed by and stored in forests, soils, and oceans. One option for slowing the rise of greenhouse gas concentrations in the atmosphere, and thus possible climate change, is to increase the amount of carbon removed by and stored in forests. As Congress debates climate change and options for addressing the issue, ideas for increasing carbon sequestration in forests are likely to be discussed.

This report examines basic questions concerning carbon sequestration in forests. The first section provides a brief background on congressional interest in forest carbon sequestration. The second describes the basic carbon cycle in forests, with an overview of how carbon cycling and storage vary among different types of forests. The third section then addresses how forest carbon is considered in the global climate change debate. This third section begins with an overview of accounting for forest carbon, then discusses the carbon consequences of forest management practices, the effects of changes in land use, and "leakage." The section then concludes with a summary of existing federal programs that could affect forest carbon sequestration.

Chapter 2 - Biochar is charcoal (similar to chars generated by forest fires) that is made for incorporation into soils to increase soil fertility while providing natural carbon sequestration. The incorporation of biochar into soils can preserve and enrich soils and also slow the rate at which climate change is affecting our planet. Studies on biochar, such as those cited in this Fact Sheet, are applicable to both fire science and soil science.

When plant materials—such as wood, bark, corn stalks, switchgrass, and the like, which scientists generically term biomass—are heated in a low- or no-oxygen atmosphere (this process is termed pyrolysis), the resulting product is called biochar. For example, pyrolysis of wood yields a familiar biochar: charcoal, once the staple fuel for backyard cookouts.

Carbon in the form of carbon dioxide (CO_2), which is released into the atmosphere when carbon-based fuels are burned and when biomass decays, is a major contributor to global climate change. For this reason, carbon sequestration (the process of capturing CO_2 before it escapes into the atmosphere) has the potential to slow global warming. Increasingly, scientists are looking at biochars as additives that can simultaneously enrich soil and sequester carbon that would otherwise be released into the atmosphere if biomass were simply left aboveground to decay.

Chapter 3 - Biochar is a charcoal produced under high temperatures using crop residues, animal manure, or any type of organic waste material. Depending on the feedstock, biochar may look similar to potting soil or to a charred substance. The combined production and use of biochar is considered a carbon-negative process, meaning that it removes carbon from the atmosphere.

Biochar has multiple potential environmental benefits, foremost the potential to sequester carbon in the soil for hundreds to thousands of years at an estimate. Studies suggest that crop yields can increase as a result of applying biochar as a soil amendment. Some contend that biochar has value as an immediate climate change mitigation strategy. Scientific experiments suggest that greenhouse gas emissions are reduced significantly with biochar application to crop fields.

Obstacles that may stall rapid adoption of biochar production systems include technology costs, system operation and maintenance, feedstock availability, and biochar handling. Biochar research and development is in its infancy. Nevertheless, interest in biochar as a multifaceted solution to agricultural and natural resource issues is growing at a rapid pace both nationally and internationally.

Past Congresses have proposed numerous climate change bills, many of which do not directly address mitigation and adaptation technologies at developmental stages, such as biochar. However, biochar may equip agricultural and forestry producers with numerous revenue-generating products: carbon offsets, soil amendments, and energy.

This report briefly describes biochar, some of its potential advantages and disadvantages, legislative support, and research and development activities underway in the United States.

Chapter 4 - Bioenergy production from forest biomass offers a unique solution to reduce wildfire hazard fuel while producing a useful source of renewable energy. However, biomass removals raise concerns about reducing soil carbon and altering forest site productivity. Biochar additions have been suggested as a way to mitigate soil carbon loss and cycle nutrients back into forestry sites; yet, little is known about the effects of intentional biochar amendments to temperate forest soil in conjunction with biomass removals for bioenergy production. In this review, the authors evaluate the potential for mobile bioenergy systems and the environmental implications of biochar application in forests. Using forest biomass that accumulates annually during forest harvest operations, bioenergy can be produced on-site and the biochar that is generated can be redistributed to return nutrients and help improve water holding capacity of the site. Little is known about the short- and long-term impacts of biochar application in forest ecosystems. Some sites may benefit from biochar application, while others show no or negative responses. Field studies on soil and vegetation responses combined with laboratory studies will elucidate the best sites for biochar application and sustainable bioenergy production.

Chapter 5 - Biopower—a form of renewable energy—is the generation of electric power from biomass feedstocks. Biopower, which comprised about 1% of electricity generation in 2008, may reduce greenhouse gas emissions, provide energy security, and promote economic development. A large range of feedstocks can be used, from woody and herbaceous biomass to agricultural residues. Each feedstock has technical and economic advantages and challenges compared to fossil fuels.

Unlike wind or solar energy, a biopower plant is considered to be a baseload power source because some biomass feedstocks can be used for continuous power production. However, ensuring a sustainable supply of biomass feedstocks is a major challenge. Although there are multiple biopower technologies, few of them except combustion have been deployed at commercial scale nationwide.

Federal policymakers are supporting biopower through feedstock supply analysis and biopower technology assessments. However, there is limited comprehensive data about the type and amount of biomass feedstock available to meet U.S. biopower needs at a national level. If the use of dedicated biomass feedstocks to generate biopower were to develop into a sizeable industry, concerns would likely include the effect of the industry on land use (i.e., how much land would it take to grow the crops needed to fuel or co-fuel power plants) and the effect on the broader economy, including farm income and food prices. To date, these have not been issues: most existing biomass feedstocks have been waste products generated by the forest products industry or by farms, or municipal solid waste for which combustion served as both a disposal method and a source of energy.

Growing crops for use as a power source would be different from using waste. Under generally accepted assumptions regarding crop yields and energy content, approximately 31 million acres— roughly the amount of land in farms in Iowa—would be needed to supply enough biomass feedstock to satisfy 6% of total 2008 U.S. electricity retail sales. When added to the amount of land needed to meet the requirements of the Renewable Fuel Standard (RFS), a federally mandated transportation fuel requirement, the potential impacts could be significant: the RFS already consumes 35% of the nation's corn crop, and its requirements will triple between 2010 and 2022 (although much of this fuel will come from feedstocks other than corn).

Beyond land use and economic impacts, others are concerned that the use of biomass feedstocks to generate biopower, particularly through combustion, could add to greenhouse gas (GHG) emission levels and exacerbate climate change concerns. They fear that certain areas may be unsustainably harvested to meet biomass feedstock demand, or that less biomass may be left for other purposes (e.g., wood and paper products). The concerns exist partly because biomass used for biopower does not face the same constraints as biomass used for liquid transportation fuels under the RFS. In addition, the idea that biomass combustion is carbon-neutral is under scrutiny. The Environmental Protection Agency has not exempted biomass combustion emissions from the Prevention of Significant Deterioration and Title V Greenhouse Gas Tailoring Rule, although it announced plans to defer for three years GHG permitting requirements for carbon dioxide emissions from biomass-fired and other biogenic sources (those produced by living organisms). In the PSD and Title V Permitting Guidance for Greenhouse Gases issued in 2010, EPA noted that biomass could be considered a best available control technology (BACT)—a pollution control standard mandated by the Clean Air Act.

Chapter 6 - Congress has expressed interest in biopower—electricity generated from biomass. Biopower, a baseload power source, has the potential to strengthen rural economies, enhance energy security, and improve the environment, proponents say. Biopower could be produced from a large range of biomass feedstocks nationwide (e.g., urban, agricultural, and forestry wastes and residues). One challenge to biopower production is a readily available feedstock supply. At present, biopower requires tax incentives to be competitive with conventional fossil fuels. If Congress considers a renewable electricity standard or other measures (e.g., farm bill energy programs) that include biopower, there may be concerns about the carbon neutrality of biopower. Congressional support for biopower has aimed to promote energy diversity and improve energy security, and has generally assumed that biopower is carbon neutral. An energy production activity is typically classified as carbon neutral if it produces no net increase in greenhouse gas (GHG) emissions on a life-cycle basis. The premise that biopower is carbon neutral has come under scrutiny as its potential to help meet U.S. energy demands and reduce U.S. greenhouse gas emissions is more closely examined.

Whether biopower is carbon neutral depends on many factors, including the definition of carbon neutrality, the feedstock type, the technology used, and the time frame examined. Carbon flux (emission and sequestration) varies at each phase of the biopower pathway, given site- and operation-specific factors. A life-cycle assessment (LCA) is a common technique to calculate the environmental footprint, including the carbon flux, of a particular biopower pathway. However, past legislation has not required a standardized LCA.

Interest in the carbon classification of biopower is in part due to sustainability and air quality concerns. Where the feedstock supply for biopower originates, if it is managed in a sustainable manner, and whether the associated air quality impacts from biopower generation are tolerable are questions that are part of the biopower carbon-neutrality debate. Congress may decide whether the current carbon-neutral designation for biopower is accurate, or whether additional carbon accounting for biopower is warranted and what impact this accounting might have on renewable energy, agricultural, and environmental legislative goals.

Rulings by the U.S. Environmental Protection Agency have raised questions about the carbon neutrality of biopower. For instance, the 2010 Prevention of Significant Deterioration and Title V Greenhouse Gas Tailoring Rule did not exempt emissions from biomass combustion. Some view EPA's decision as equating biomass emissions with fossil fuel emissions. EPA decided in 2011 to defer for three years GHG permitting requirements for carbon dioxide emissions from bioenergy and other biogenic stationary sources in order to conduct a detailed examination of the science associated with these emissions. EPA's Scientific Advisory Board conducted an independent review of the agency's biogenic accounting framework and released its findings in September 2012. The board acknowledged the "daunting task" of assessing the greenhouse gas implications of bioenergy, and the "narrow regulatory boundaries" within EPA's purview that limit the consideration of greenhouse gas flux at various points along the bioenergy pathway.

State perspectives on the tailoring rule are divided. Some states contend that treating biomass combustion the same as fossil fuel combustion will result in excessive permitting requirements and fees that jeopardize renewable energy development. Other states argue that not treating it the same will aggravate climate change over time.

In: Carbon Considerations
Editors: W. Green and L. Wayman

ISBN 978-1-62618-435-0

Chapter 1

CARBON SEQUESTRATION IN FORESTS*

Ross W. Gorte

SUMMARY

Widespread concern about global climate change has led to interest in reducing emissions of carbon dioxide (CO_2) and, under certain circumstances, in counting additional carbon absorbed in soils and vegetation as part of the emissions reductions. Congress may consider options to increase the carbon stored (sequestered) in forests as it debates this and related issues.

Forests are a significant part of the global carbon cycle. Plants use sunlight to convert CO_2, water, and nutrients into sugars and carbohydrates, which accumulate in leaves, twigs, stems, and roots. Plants also respire, releasing CO_2. Plants eventually die, releasing their stored carbon to the atmosphere quickly or to the soil where it decomposes slowly and increases soil carbon levels. However, little information exists on the processes and diverse rates of soil carbon change.

How to account for changes in forest carbon has been contentious. Land use changes—especially afforestation and deforestation—can have major impacts on carbon storage. Foresters often cut some vegetation to enhance growth of desired trees. Enhanced growth stores more carbon, but the cut vegetation releases CO_2; the net effect depends on many

* This is an edited, reformatted and augmented version of a Congressional Research Service publication, RL31432, dated August 6, 2009.

factors, such as prior and subsequent growth rates and the quantity and disposal of cut vegetation. Rising atmospheric CO_2 may stimulate tree growth, but limited availability of other nutrients may constrain that growth.

In this context, timber harvesting is an especially controversial forestry practice. Some argue that the carbon released by cutting exceeds the carbon stored in wood products and in tree growth by new forests. Others counter that old-growth forests store little or no additional carbon, and that new forest growth and efficient wood use can increase net carbon storage. The impacts vary widely, and depend on many factors, including soil impacts, treatment of residual forest biomass, proportion of carbon removed from the site, and duration and disposal of the products. To date, the quantitative relationships between these factors and net carbon storage have not been established.

Some observers are concerned that "leakage" will undermine any U.S. efforts to sequester carbon by protecting domestic forests. By leakage, they mean that wood supply might shift to other sites, including other countries, exacerbating global climate change and causing other environmental problems, or that wood products might be replaced by other products that use more energy to manufacture (thus releasing more CO_2). Others counter that the "leakage" arguments ignore the enormous disparity in ecological systems and product preferences, and discount possible technological solutions.

Several federal government programs affect forestry practices and thus carbon sequestration. Activities in federal forests affect carbon storage and release; timber harvesting is the most controversial such activity. Federal programs also provide technical and financial help for managing and protecting private forests, and tax provisions affect private forest management. Various federal programs can also affect the extent of forested area, by supporting development (which may cause deforestation) or encouraging tree planting in open areas, such as pastures.

Global climate change is a widespread and growing concern that has led to extensive international discussions and negotiations.[1] Responses to this concern have focused on reducing emissions of greenhouse gases, especially carbon dioxide, and on measuring carbon absorbed by and stored in forests, soils, and oceans. One option for slowing the rise of greenhouse gas concentrations in the atmosphere, and thus possible climate change, is to increase the amount of carbon removed by and stored in forests. As Congress debates climate change and options for addressing the issue, ideas for increasing carbon sequestration in forests are likely to be discussed.

This report examines basic questions concerning carbon sequestration in forests. The first section provides a brief background on congressional interest in forest carbon sequestration. The second describes the basic carbon cycle in forests, with an overview of how carbon cycling and storage vary among different types of forests. The third section then addresses how forest carbon is considered in the global climate change debate. This third section begins with an overview of accounting for forest carbon, then discusses the carbon consequences of forest management practices, the effects of changes in land use, and "leakage." The section then concludes with a summary of existing federal programs that could affect forest carbon sequestration.[2]

BACKGROUND: CONGRESSIONAL INTEREST IN CARBON SEQUESTRATION

The widespread and growing concern over global climate change has led to extensive international negotiations. In 1992, at the Earth Summit in Rio de Janeiro, the United Nations Framework Convention on Climate Change (which included voluntary pledges to reduce greenhouse gas emissions) was opened for signature. President George H. W. Bush signed this treaty, which was then ratified by the U.S. Senate.

Subsequent negotiations led to the 1997 Kyoto Protocol, under which the developed nations agreed to specified reductions in their emissions of greenhouse gases. President Clinton signed the Kyoto Protocol, but did not submit it to the Senate for ratification. Early in 2001, the George W. Bush Administration decided to reject the Kyoto Protocol, and withdrew from active participation in negotiations on the issues that remain to be resolved.[3] Although enough other parties have ratified the Protocol to bring it into force, the lack of U.S. involvement means that the United States will not participate in the emissions trading or other elements of the Kyoto Protocol activities that might relate to carbon sequestration.

The most voluminous greenhouse gas produced by humans is carbon dioxide (CO_2). In calculating overall carbon emissions, the Protocol allows certain removals of carbon by a nation's forests and soils—"carbon sinks"—to be counted and deducted from emissions. Thus, one option for mitigating greenhouse gas emissions—and thus possible climate change—is to increase the amount of carbon stored in forests.

Carbon sequestration, and the extent to which it can be counted as a reduction in a nation's carbon emissions, have been the focus of substantial controversy in international negotiations subsequent to the Kyoto Protocol.[4] The United States, with its extensive forests, argued that the carbon absorbed by them should be allowed to offset emissions, with no quantitative limit to the amounts that can be counted in this way. The European Union argued strongly in negotiations prior to 2001 that there should be fairly strict limits on how much carbon absorbed by "sinks" such as forests could be counted against emissions. In final negotiations during 2001 on rules to implement the Kyoto Protocol, after the United States had withdrawn from the negotiations, a compromise was reached that allows significant credit for carbon sinks (removals and storage of carbon). The Members of Congress attending the negotiations prior to 2001 followed this debate with interest, and were aware of the potential impacts of the various possible results of the negotiations. In particular, if emissions trading were to begin under the Kyoto Protocol, forest owners and managers in countries that were parties to the treaty might be able to receive credits and participate in the trading regime.

Administration and congressional interest in carbon sequestration continues, but U.S. participation in the Kyoto process is moot at this time. It is not clear whether a domestic forest carbon program might be established, although options have been discussed in legislative proposals. Protecting forests in developing countries, which might earn credits under the Kyoto Protocol, is already supported under the Tropical Forest Conservation Act (P.L. 105-2 14; 22 U.S.C. §§2341, et seq.).[5]

Mitigating climate change by enhancing forest carbon sequestration may be a relatively low-cost option and would likely yield other environmental benefits. However, forest carbon sequestration faces challenges: measuring the *additional* carbon stored (over and above what would occur naturally); monitoring and verifying the results; and preventing leakage. Numerous issues regarding the carbon cycle in forests, monitoring the levels and changes in forest carbon, and the scientific uncertainties about the relationships among forests, carbon, and climate change are likely to be the subject of ongoing federal research efforts, with funding and oversight by the Congress.

CARBON CYCLING IN FORESTS

Photosynthesis is the chemical process by which plants use sunlight to convert nutrients into sugars and carbohydrates. Carbon dioxide (CO_2) is one of the nutrients essential to building the organic chemicals that comprise leaves, roots, and stems. All parts of a plant—the stem, limbs and leaves, and roots—contain carbon, but the proportion in each part varies enormously, depending on the plant species and the individual specimen's age and growth pattern. Nonetheless, as more photosynthesis occurs, more CO_2 is converted into biomass, reducing carbon in the atmosphere and sequestering (storing) it in plant tissue (vegetation) above and below ground.

Plants also respire, using oxygen to maintain life and emitting CO_2 in the process. At times (e.g., at night and during winter seasons in non-tropical climates), living, growing forests are net emitters of CO_2, although they are generally net carbon sinks over the life of the forest.

When vegetation dies, carbon is released to the atmosphere. This can occur quickly, as in a fire,[6] or slowly, as fallen trees, leaves, and other detritus decompose. For herbaceous plants, the aboveground biomass dies annually and begins to decompose right away, but for woody plants, some of the aboveground biomass continues to store carbon until the plant dies and decomposes. This is the essence of the carbon cycle in forests—net carbon accumulation (sequestration) with vegetative growth, and release of carbon when the vegetation dies. Thus, the amount of carbon sequestered in a forest is constantly changing with growth, death, and decomposition of vegetation.

In addition to being sequestered in vegetation, carbon is also sequestered in forest soils. Carbon is the organic content of the soil, generally in the partially decomposed vegetation (humus) on the surface and in the upper soil layers, in the organisms that decompose vegetation (decomposers), and in the fine roots.[7] The amount of carbon in soils varies widely, depending on the environment and the history of the site. Soil carbon accumulates as dead vegetation is added to the surface and decomposers respond. Carbon is also "injected" into the soil as roots grow (root biomass increases). Soil carbon is also slowly released to the atmosphere as the vegetation decomposes. Scientific understanding of the rates of soil carbon accumulation and decomposition is currently not sufficient for predicting changes in the amount of carbon sequestered in forest soils.

The Forest Cycle

Forests generally go through cycles of growth and death, sequestering and releasing carbon. Some forests begin on spacious sites, with little or no existing vegetation, that may have been cleared by a natural disaster (most commonly wildfire) or by human activities (e.g., for agriculture). Other forests are relatively continuous, with natural clearings typically limited to the area occupied by one or a few large trees killed by lightning, disease, and such.

Regardless of the size or origin of a clearing, most forests begin from essentially bare land, with some carbon stored in the soil (how much depends on the environment and history of the site, especially the last clearing process).

As trees and other woody plants become established, carbon stored on the site increases as woody biomass increases and as annual vegetation (e.g., tree leaves and herbaceous plants) typically grows faster than it decomposes. Productivity for commercially usable wood generally follows an S-shaped curve, with the volume growing at an increasing rate for many years, to a point known to foresters as the culmination of mean annual increment (generally taking 20 to 100 years or more, depending on the fertility of the site and the tree species), and then growing at a decreasing rate for many more years. In theory, forests can eventually become "over-mature," where the loss of commercial volume through tree mortality equals or exceeds the additional growth on the remaining trees. However, one study has shown that some old-growth ("over-mature") forests continue to accumulate carbon in their soils.[8]

The relationship between commercially usable wood produced and carbon sequestered varies substantially in three ways. First, the proportion of carbon in a tree's commercial wood (compared to the noncommercial biomass in bark, limbs, roots, and leaves or needles) varies among species; some (e.g., pines and other conifers) have a greater proportion of their total carbon in commercial wood.[9] Second, the proportion of carbon in a tree's commercial wood undoubtedly changes over time; while a temporal graph of carbon storage is probably also S- shaped (as for commercial wood productivity), the changes in timing and rates of increase (that cause the characteristic S shape) almost certainly differ. Finally, a significant portion of the vegetative carbon sequestered in a forest is in other plants—noncommercial species of trees, shrubs, grasses, and other herbaceous plants. The amount of carbon stored in this other (noncommercial) growth varies widely among forests. Thus, although many research studies assume a fixed relationship between commercial wood inventories and the amount of carbon stored,[10] the traditional measures of commercial wood production might not be very accurate for estimating carbon sequestration in forests.

Eventually, trees die. They may be cut down, burned in a wildfire, blown over or snapped off in a wind or ice storm, or killed by insects or diseases. Death can happen to a single tree in a forest, creating a small opening, or to all or most trees in an area. How quickly the carbon is released to the atmosphere depends on the cause of tree death, on whether it is harvested for use, and on various environmental factors. As noted above, fires quickly break down biomass and release an enormous amount of CO_2 into the atmosphere. Natural death and decay may require several weeks to several decades to completely decompose the biomass (depending on site conditions), putting some carbon into the soil and some directly into the atmosphere. Timber harvesting can store some vegetative carbon for very long periods in solid wood products with long-term uses (e.g., construction lumber in houses), while tree tops and limbs and noncommercial species are left to decay or to be burned. These possibilities are discussed in more depth below, under "Forestry Events and Management Activities."

Forest Types

Carbon sequestration and release vary substantially by forest. Nonetheless, some generalizations are possible, because of the relative similarity of forests in specific "biomes"[11]—tropical, temperate, and boreal forests. **Table 1** shows average carbon levels sequestered in vegetation and soils for several major biomes, and the weighted average for all biomes.[12]

Table 1. Average Carbon Stocks for Various Biomes (in Tons Per Acre)

Biome	Plants	Soil	Total
Tropical forests	54	55	109
Temperate forests	25	43	68
Boreal forests	29	153	182
Tundra	3	57	60
Croplands	1	36	37
Tropical savannas	13	52	65
Temp. grasslands	3	105	108
Desert/semidesert	1	19	20
Wetlands	19	287	306
Weighted Average	***14***	***59***	***73***

Source: Adapted from Intergovernmental Panel on Climate Change, "Table 1: Global carbon stocks in vegetation and carbon pools down to a depth of 1 m [meter]," *Summary for Policymakers: Land Use, Land-Use Change, and Forestry. A Special Report of the Intergovernmental Panel on Climate Change*, at *http:// www.ipcc.ch/pub/* srlulucf-e.pdf, p. 4.

Tropical Forests

Tropical forests are generally defined by their location—between the Tropic of Cancer and the Tropic of Capricorn (23° north and south of the Equator, respectively). Some tropical forests are relatively dry, open woodlands, but many receive heavy rains and are called moist or humid tropical forests; these are the classic rainforests, or "jungles." Tropical forests contain an enormous diversity of "hardwood" tree species,[13] and are difficult to categorize into "forest types," because of this breadth of species diversity.

Moist tropical forests are important for carbon sequestration, because they typically have high carbon contents—averaging nearly 110 tons per acre. (See **Table 1**.) About half of the carbon in moist tropical forests is contained in the vegetation, a higher percentage and a much higher quantity than in any other biome. The remaining carbon is in tropical forest soils. Tropical forest soils have only modest carbon levels (compared with other biomes), because the dead biomass rapidly decomposes in the warm, humid conditions and the minerals rapidly leach out of tropical forest soils.

Temperate Forests

Temperate forests typically occur in the mid-latitudes—generally to about 50° north and south of the Equator (a little farther north in Europe, because of the continental warming from the Gulf Stream). There are a large variety of temperate forests, including hardwood types (e.g., oak- hickory and maple-beech-birch), softwood types (e.g., southern pine, Douglas-fir, and lodgepole pine), and a few mixed types (e.g., oak-pine). However, within each forest type, and overall, temperate forests have much lower tree species diversity than tropical forests.

Temperate forests generally contain less carbon than tropical forests, averaging nearly 70 tons per acre. More than one-third of the carbon is stored in the vegetation, and nearly two-thirds in the soil. The higher proportion (but lower level) in the temperate forest soils (compared to tropical forest soils) is because of slower decomposition rates. Many of these forests are managed to produce commercial wood products, and the management practices used in temperate forests can thus have a significant impact on carbon sequestration.

Boreal Forests

Boreal forests generally occur north of temperate forests, mostly in Alaska, Canada, Scandinavia, and Russia. (The only boreal forests in the Southern Hemisphere are on mountaintops in New Zealand and high in the

Andes Mountains of South America.) Boreal forests are dominated by conifers—mostly spruce, fir, and larch, with scattered birch and aspen stands.

Boreal forests generally contain more carbon than temperate or tropical forests, averaging more than 180 tons per acre. Less than one-sixth of boreal forest carbon is in vegetation. The rest, 84%, is in boreal forest soils—about three times the amount in temperate and tropical forests, and far higher than any other biome except wetlands.[14] Carbon accumulates to high levels in boreal forest soils because of the very slow decomposition rates, owing to the short summers and high acidity of conifer forest soils, both of which inhibit decomposition. The high boreal forest soil carbon level is important for carbon cycling, because many believe that management activities that disturb boreal forest soils can increase their release of carbon.

MEASURING AND ALTERING FOREST CARBON LEVELS

Aside from the questions of whether climate change is occurring and whether human activities are the cause, the role of forestry and land use in mitigating climate change has been quite controversial. The disputes are largely the result of the scientific uncertainties in measuring changing carbon levels in forests, changing land uses, and changing demand for products. This section summarizes forest carbon accounting concerns, possible consequences of changes in land use and of forest management events and practices, "leakage," and existing federal programs related to these concerns.

Forest Carbon Accounting

Different countries have various views on how to count carbon sequestered or released from forests. In general, countries with extensive and expanding forests (e.g., Russia, Canada, Brazil, and the United States) prefer a full accounting, while countries with less forestland (e.g., many European countries) are concerned about the potential to overstate the carbon benefits of forestry management practices and land use changes that enhance carbon sequestration. Countries with net deforestation rates are also concerned about counting forest sequestration, because it could effectively increase their net emission rated under international agreements.

Article 3.3 of the Kyoto Protocol allows counting the carbon effects (both sequestration and release) of afforestation, reforestation, deforestation, and other forestry and land use changes that have occurred since 1990, if the change in carbon stock is verified. Verification requires a system for estimating the carbon effects—because a census of carbon changes on every forested acre is infeasible—and for reporting the carbon effects.

For countries with carbon commitments (rather than for projects), the surest, easiest system for verifying the change in carbon levels is to measure the change in the levels from the beginning to the end of the relevant time period—1990 (the baseline) and 2008-2012 (the Kyoto Protocol commitment period); however, this is a very slow and expensive approach.[15] A variety of models can be used for estimating carbon level changes. The two basic approaches are:

- a "land-based" approach, which begins by identifying the acceptable activities for sequestering carbon and estimating the carbon consequences of those activities, and then monitors the *lands* to determine the extent to which those activities occur; and
- an "activity-based" approach, which also begins by identifying the acceptable activities for sequestering carbon and estimating the carbon consequences of those activities, and then monitors the *activities* to determine the extent to which those activities occur.[16]

The approach taken affects the intensity and spatial scale of the monitoring required, and different models impose different requirements for data, boundary conditions, carbon stocks, and more. However, the Intergovernmental Panel on Climate Change contends that, regardless of the approach and model:[17]

> A well-designed carbon accounting system would provide transparent, consistent, comparable, complete, verifiable, and efficient recording and reporting of changes in carbon stocks and/or changes in greenhouse gas emissions by sources and removals by sinks from applicable land use, land-use change, and forestry activities and projects under relevant Articles of the Kyoto Protocol.

There are significant difficulties in achieving such a "well-designed carbon accounting system."[18] Some observers have noted that the "language, terminology, and accounting methods contained in the [Kyoto Protocol] are somewhat vague and can be interpreted in different ways."[19] In addition, there are scientific uncertainties in measuring the 1990 baseline carbon stocks, the lands treated, the carbon impacts of various treatments, and the question of "leakage," as discussed in the following sections.

Land Use Changes

Over time, forests can grow on lands that were in other uses (e.g., croplands), and vice versa. Deforestation is the conversion of forests to pasture, cropland, urban areas, or other landscapes that have few or no trees. Afforestation is planting trees on lands that have not grown trees in recent years, such as abandoned cropland.

Not all land use changes have comparable carbon consequences. Tropical deforestation, if a pasture replaces the forest, could release carbon and substantially reduce carbon sequestration, since tropical savannahs store less than half as much carbon on average as tropical forests.[20] In contrast, in temperate zones, forest conversion to native grasslands would likely release carbon but might actually *increase* carbon sequestration, since temperate grasslands store 50% more carbon on average than do temperate forests. If, however, the temperate deforestation is for crop production, carbon would be released and less carbon would be sequestered, since croplands store only half as much carbon on average as temperate forests.[21] The actual impacts—both the quantity and the timing or duration of the change—probably depend on the actual practices employed on the various sites. Other changes may be less obvious. Converting a "working forest" (i.e., one that is managed to produce timber and other values) to a rural residential subdivision would likely reduce forest cover and perhaps undergrowth (thus releasing some vegetative carbon), but the remaining vegetation would likely continue to grow, and thus may continue to sequester carbon for longer in the subdivision than in a working forest (depending on the nature and duration of the wood products derived from the working forest). Thus, it may not be possible to draw simple conclusions about the carbon consequences of land use changes, especially in temperate zones.

According to the *1997 National Resources Inventory*, more than 25 million acres of private forestland in the United States were converted to other uses between 1982 and 1997.[22] Of these, nearly 12 million acres were developed for other uses, while 4 million acres were converted to pasture and nearly 4 million acres were either inundated by water (deforested) or transferred to federal ownership (and probably remained forested). During the same period, nearly 26 million acres were converted to private forestland—a net increase of nearly a million acres of private forestland. Of these, more than half (nearly 14 million acres) were previously pastureland, and more than 5 million acres were cropland.

The conversion of forestland to other uses is dominated by development. The rate of development reflects economic growth, interest rates, and development incentives—stronger growth, lower interest rates, greater incentives stimulate more development (and more conversion of forestland), while slower growth or recession, higher interest rates, or weaker incentives may retard development. While federal monetary, fiscal, and tax policies clearly affect development rates, their impact on forestland conversion is likely less of a consideration than the goal of trying to provide stable economic growth for the U.S. economy. The conversion of pasture and cropland to forests is also affected by the general economy, but is also likely to be directly affected by existing federal programs, such as the forestry assistance programs and other agricultural conservation programs (discussed below).

Forestry Events and Management Activities

As discussed above, forests cycle carbon, accumulating carbon from the atmosphere during some periods and releasing carbon to the atmosphere at other times. When forest practices alter the vegetation on a site, they alter this ebb and flow of carbon storage and release by changing biomass levels, vegetative growth patterns, and soil structure and composition. There have been debates in the literature, and at least one congressional hearing, about the carbon impacts of various forest management practices.[23] This section examines the implications of various forestry practices and events for carbon sequestration. First, however, it examines generic considerations of what happens to the carbon in cut vegetation and in soils from forestry practices.

Vegetation and Soil Carbon

A wide array of practices are used in managing forests. Most involve altering the vegetation, and thus the carbon cycle, on the site. Before discussing specific practices, it is important to examine what happens to vegetation that is cut but not removed from the forest and what happens to the soil carbon, as these factors apply to most, if not all, forest management practices.

What Happens to the Cut Vegetation?

Most forest practices involve cutting some of the existing vegetation on a site—to harvest commercial wood, to focus growth on fewer trees, to reduce competition among plants, and more. What happens to the cut vegetation is critical for assessing carbon sequestration and release.

When commercial harvests are involved, some of the biomass is removed from the site and turned into products. (The carbon consequences of removed vegetation are discussed below.) How much biomass (and carbon) is removed has not been broadly quantified; various studies report that 50% to 80% of tree carbon (excluding forest soil carbon) is removed in commercial timber harvest operations.[24] Others have stated that "wood products formed only a modest fraction of the total" carbon stored by a forest.[25] This is consistent with a newer report showing wood products (and landfills) accounting for a little more than 6% of forest-related carbon stocks.[26] The proportion of carbon and biomass removed from any particular site varies widely, and depends on the species involved, the density of the stand (which affects both tree form and undergrowth vegetation), the diversity of tree species and tree sizes in the stand, and various environmental factors (e.g., the site's climate and soil fertility).

Much biomass (and carbon) remains on the site; for some forest practices, such as precommercial thinning (described below), all the cut biomass remains on the site. The remaining biomass— coarse roots, limbs, leaves or needles, and other unusable woody material, often called "slash" in timber harvesting—begins to deteriorate, and to release carbon. How fast the carbon is released depends on whether and how the biomass is treated. Slash treatments include rolling, chopping, and crushing, all of which are designed to compact the biomass and accelerate its deterioration, which typically takes several weeks to several years, depending substantially on the climate. Burning is also common, and leads to deterioration (and release of some of the carbon to the atmosphere) in minutes or hours, rather than weeks, months, or years. Many research studies presume that the carbon from slash is released within a year of the harvest; others presume that it is quickly absorbed by new growth resulting from the treatment. Data on the extent of various slash treatments and on the carbon impacts of the various treatments are sparse.

What Happens to Forest Soil Carbon?

Relatively little literature on forests and carbon addresses the mechanisms by which carbon is accumulated in and released by forest soils. One source noted "the large uncertainty in estimates of change in soil carbon."[27] Dead vegetation is broken down by decomposers—primarily invertebrate animals and fungi. Decomposition releases carbon to the atmosphere and incorporates carbon as organic matter (humus) in the soil. The dead carbon in the soil (i.e., not the decomposers or fine plant roots, which comprise the living carbon in the soil) slowly continues to decompose, changing from organic to inorganic forms; the inorganic forms eventually (1) dissolve in percolating rainwater and leach through the soil into groundwater and surface waters; (2) oxidize and get released directly to the atmosphere; or (3) are absorbed by new vegetation. The rates at which carbon from dead vegetation returns to the atmosphere likely depend on a variety of factors, such as the nature of the vegetation, the composition of the soil, and the humidity levels, all of which affect the type and quantity of decomposers on a site.

Activities that disturb soils almost certainly decrease soil carbon levels. Some studies presume a loss of soil carbon following a soil-disturbing activity, such as timber harvesting.[28] Others note the loss without quantifying it.[29] In general, disturbing soils accelerates decomposition rates, by increasing exposure of soil carbon to air, thus accelerating oxidation. Activities that disturb soils also kill living soil carbon—invertebrate animals, fungi, and fine roots—which then begin to decompose. Thus, forestry activities that disturb soils, particularly activities to remove the cut vegetation (such as commercial timber harvesting), will also likely reduce soil carbon levels. Forestry activities that do not disturb soils, such as fertilization and prescribed burning, probably have much less effect on soil carbon, although they affect the forest carbon balance in other ways (as discussed below).

Forest Events — Wildfires[30]

Fires in forests and grasslands are common events that significantly affect the carbon cycle. As noted above, fire is a self-sustaining chemical process that quickly mineralizes organic matter; in minutes, fires convert organic matter into its components—minerals, water vapor, and CO_2. Wildfires are a very significant source of atmospheric CO_2, and the need to control wildfire to reduce CO_2 emissions, as well as for safety reasons, has been discussed. Furthermore, the likelihood, extent, and/or frequency of wildfires may be exacerbated by expected climate change.

Wildfire is a natural phenomenon, although efforts are made to manage wildfires. One study estimated that in pre-industrial America (200-500 years ago), wildfires burned nearly 10 times as many acres and as much biomass as in recent years.[31] Especially in temperate ecosystems, with dry biomass and lightning, wildfires are inevitable.

Some have proposed wildfire protection by reducing biomass levels in forests. Several of the forestry practices discussed below are intended at least partly to reduce wildfire risks by reducing biomass levels—wildfire fuels. The presumption is that lower fuel loads will lead to fewer and less intense fires that are easier (and less costly) to control. Numerous on-the-ground anecdotes support this presumption. However, little empirical research documents this presumption. One study found that "scant information exists on fuel treatment efficacy for reducing wildfire severity."[32] A summary of wildfire research reported that prescribed burning generally reduced fire severity, that mechanical fuel reduction did not consistently reduce fire severity, and that little research has examined the potential impacts of mechanical fuel reduction with prescribed burning or of commercial logging.[33] The possibility of removing some noncommercial biomass from the forest and using it to produce energy is also being considered, and is discussed below.

One study found that fires in boreal forests may actually cool the atmosphere, despite their large CO_2 releases.[34] The researchers found that loss of dark-colored conifers increased snow cover and that subsequent regrowth was of light colored birch and aspen. The result was significantly greater *albedo* (light reflectivity), which led to greater cooling than the CO_2 release led to warming. The extent to which this conclusion can be extended is unclear, and of course, boreal forest fires may have other undesirable effects.

Forestry Practices

In general, forestry practices are used for four basic goals or purposes: to establish trees on a site; to reduce vegetative competition; to improve tree growth in other ways; and to harvest the commercially usable wood. In the climate change negotiations and legislation, credit for carbon sequestered by forests is intended to be allowed only for *additional* carbon sequestered because of changes in forestry practices; carbon sequestered by normal, business-as-usual practices is not additional, and usually cannot be included for carbon sequestration activities.

Stand Establishment

One basic objective of forestry is getting trees to start growing. On sites that have recently had trees on them, but are now cleared (because of timber harvesting or natural disaster), the practice is called reforestation. Tree planting on sites that have not recently had trees on them, such as pastures, is called *afforestation*. In addition, *interplanting* (planting additional trees) is used on sites that have fewer trees than are considered desirable.

Reforestation can use natural or artificial methods. Natural regeneration relies on tree seeds from surrounding forest stands. Artificial regeneration includes planting seeds, or more commonly seedlings from nurseries, on the site. Advantages of natural regeneration include lower cost and greater stand diversity (both of tree species and of the genetic diversity of the dominant tree species), which generally increases forest resilience and resistance to pests and pathogens. Advantages of artificial regeneration include greater dominance of commercially desirable tree species, greater control over the number of trees established, and more rapid establishment of trees, all of which increase growth of the desired trees. Artificial regeneration may be necessary for afforestation, since surrounding tree seed sources might be inadequate or nonexistent. Artificial regeneration likely provides more commercial wood growth faster, and may sequester more carbon faster, than natural regeneration.

Establishing stands of trees will generally sequester more carbon than leaving the sites without forest cover. Savannas and other non-forest biomes store much less carbon in their vegetation, and *may* reach a plateau or stable carbon stock in their soils in only a few years. In contrast, forests continue to sequester additional carbon in vegetation and roots as it grows for many years—usually at least decades, and even centuries in some ecosystems. (Note, however, from **Table 1**, that temperate grasslands have greater average carbon stocks than temperate forests, generally because perennial grasses increase soil carbon more than forests.) Thus, reforestation and especially afforestation, regardless of whether by natural or artificial regeneration, generally provide continuing additional carbon sequestration for an extended period.

Reducing Competition

Another basic forestry practice is to encourage growth of the commercially desired trees by killing other vegetation that competes with the trees for space, light, and nutrients. This practice is called *release* when the competing vegetation is as tall as or taller than the desired trees, or when it is of a different species (e.g., palmetto growing under southern pine trees). Release can be done chemically (with herbicides) or mechanically (with machines or tools) to kill the unwanted vegetation.

Thinning is the forestry practice of removing some of the desired tree species (as well as the undesired tree species) when the competition for space, light, and nutrients reduces the growth rate of the commercial timber volume. Precommercial thinning occurs when the trees are too small to have any commercial value (generally less than 5.5 inches in diameter), while commercial thinning is the practice of selling the trees to be removed.

Thinning and release are often proposed as forest management practices to increase forest carbon sequestration. Some models estimate total carbon on a site as a fixed relationship to commercial wood volume on a site. Since thinning and release increase commercial timber growth, these models would similarly project thinning and release to increase carbon sequestration.

Others, however, observe that the purpose of thinning and release is often to concentrate the same amount of growth on fewer stems.[35] One study examined the potential of thinning and release to increase carbon storage.[36] This study found that thinning only increased total carbon storage in dense, young stands where severe competition significantly reduced growth rates; in other cases, the practice was "redistributing stand growth to a smaller number of larger trees," with little overall change in carbon storage. The author also noted that thinning may increase the loss of soil carbon, by reducing canopy cover and disturbing the surface and thus accelerating decomposition rates. In contrast, release in a southern pine ecosystem was found to increase total carbon storage over the life of the stand, and to promote soil carbon storage. Thus, forestry practices that reduce vegetative competition apparently increase carbon sequestration in some circumstances, but not in others, limiting generalizations about their potential for increasing carbon sequestration.

Other Growth Improvement

Other forestry practices are also intended to increase tree growth rates. Pruning removes the lowest branches of a commercial tree, which may stimulate some upward growth, but generally emphasizes wood value (growing clear wood, without knots) rather than growth. It is not used much, because it has been found to be unprofitable in most situations.

Fertilizing forest stands is another practice to increase tree growth. Applying fertilizers to forests can significantly increase growth rates, if the nutrient being applied (nitrogen, potassium, phosphorus, etc.) is in short supply in the forest soil. Furthermore, fertilization is likely to stimulate all vegetative growth, not just tree growth. This is borne out in research on the impact of forestry practices on carbon sequestration.[37] An important question, whose answer is not apparent from the research, is whether the accelerated growth rate from fertilization persists for a long time (i.e., whether the growth rate remains higher for an extended period), whether it produces a short-term increase for which the pre-fertilized rate is sustained for a long time (i.e., whether the pre-fertilization growth rate is maintained after the short-term increase); or whether other factors limit long-term growth rates (i.e., whether growth rates after the short-term increase are less than the pre-fertilization growth rates, because other nutrients are overdrawn by the fertilizer-stimulated growth).

Some have suggested that greater atmospheric CO_2 levels could fertilize forests, stimulating tree growth. A number of studies artificially increased CO_2 levels in tree plantations, and found that growth rates did increase;[38] others, however, question whether the increased growth rate can be sustained.[39] At the broader scale, in at least some areas, other nutrients (especially nitrogen) are likely to limit the ability of forests to expand growth with more CO_2available.

Another forestry practice, which is becoming more widely used, is prescribed burning— intentionally setting fires in certain forest areas under specified weather and fuel conditions.[40] Prescribed burning typically produces many forest benefits, including less competing vegetation (akin to release or thinning), lower fuel loads that may contribute to catastrophic wildfires, and a flush of nutrients (since fire reduces biomass to its mineral components). However, prescribed fires also present a risk, since they occasionally escape from the prescribed areas and can cause damage; they also generate substantial amounts of carbon dioxide (one of the mineral components of biomass) and other air pollution. In the short term, prescribed fires clearly increase atmospheric carbon levels. To the extent that prescribed burning reduces catastrophic wildfires, this practice may be shifting one source of carbon emissions (wildfire) to another (prescribed fire); however, to date, no quantitative relationship has been established between prescribed burning and the extent and severity of wildfires. In addition, at least for dead biomass on a site, prescribed burning merely concentrates into a few minutes or hours the carbon release that otherwise occurs over a few weeks to a few years in other forms of biomass decomposition. Thus, it is not clear how much of the carbon release from prescribed burning is in addition to the carbon release that might otherwise occur from forests.

Timber Harvesting

For wood that has been removed from the forest, the rate of carbon release depends on what is done with the wood. For *sawtimber* (logs of at least 11.5 inches in diameter), about half is converted into solid wood products (primarily lumber and plywood); another 10% is bark, and the remaining 40% is sawdust and wood scrap.[41] Lumber and plywood have differing usable lives, depending on the use of the wood, and ranging from less than 10 years for shipping pallets to 100 years or more for residential construction.[42] Clearly, some wood—broken pallets, furniture, concrete forms, etc.—is disposed in landfills (and probably sequesters carbon there) and some is burned, but the majority of carbon in solid wood products remains sequestered in those products for decades. Most (more than 95%) of the bark and sawdust are either used as pulp to make paper or burned to produce energy (thus substituting for timber used in papermaking or for fossil fuels); less than 5% of waste wood from sawmills ends up in landfills.

For *pulpwood* (logs or bolts less than 11.5 inches in diameter and usually less than 8 feet in length) and waste from sawtimber processing, virtually all the wood fiber (the cellulose and hemi-cellulose) is used in paper products. The spent pulping liquors (the chemicals that dissolve the lignin, the "glue" which holds cellulose in a rigid structure) are generally used to produce energy.[43] Any waste paper in the production facility is generally recycled into pulp. Other than in energy production (which substitutes for fossil fuels), there is little paper waste that returns carbon directly to the atmosphere. In contrast to solid wood products, which may sequester carbon for decades, most paper products have relatively brief duration, releasing their carbon to the atmosphere relatively quickly—in less than a year for many paper products, and in less than 10 years for most paper products.[44] However, paper can also be recycled—dissolved, cleaned, and made into new paper products. Increasing the recycling of *post-consumer waste* paper (the paper thrown away by consumers and most likely to end up in landfills) can reduce the carbon released by paper production and use.

The carbon impacts of commercial timber harvesting have been debated extensively, but with little resolution. Some have calculated that harvesting timber from an "over-mature" forest can sequester substantial additional carbon, because (a) the forest is currently sequestering little additional carbon (the amount stored is large, but annual addition from tree growth is small or even negative), (b) the timber can continue to store carbon for decades in long-term solid wood products, and (c) the newly established stand can sequester large amounts of carbon through its vigorous growth.[45] Others have calculated that the carbon released by harvesting operations substantially exceeds the additional carbon sequestered by new forest stands.[46] One source has stated that timber harvesting (in a heavy thinning or selection harvest) reduces carbon storage, "because the growth of residual trees is less rapid than the decomposition of the detritus and harvested wood products."[47] Another study has shown that some old-growth forests continue to accumulate carbon in their soils.[48] All of these conclusions may be valid in certain circumstances; the consequences probably depend on a variety of factors, such as which products are manufactured, how those products are used, how much carbon is left on the site, and what happens to it. There are, of course, other considerations associated with discussions of harvesting old-growth forests. Thus, the carbon consequences of timber harvesting, or delaying or preventing timber harvesting, are uncertain, being highly variable and highly site-specific.

In addition, reduced impact logging (RIL) has been developed primarily to reduce the damage that timber harvesting can do to soils and to residual trees. It is becoming more widely discussed, especially for tropical forest harvests, but descriptions of RIL are generally either lacking in details or highly site-specific with limited general applicability. This is probably because the practices that will reduce logging damages depend on a variety of site conditions, such as soil type and water content, tree species diversity, etc. Nonetheless, as RIL becomes more widely practiced, it seems likely that logging damages (including carbon release) will decline.

Wood Energy

Using wood residues for energy production has occurred for many years at wood production facilities. The old "teepee" burners for disposing of wood waste are all defunct, and as noted above, the wood waste not used for paper production is already being used to produce energy to operate lumber and plywood mills. Even 30 years ago, less than 4% of the woody biomass removed from forests ended up as unused wood residues.[49]

Wood can be used to produce energy either by burning it directly, modifying it to produce more consistent burning characteristics (e.g., by pulverizing it and compressing it into pellets), or by digesting it to produce liquid fuel (methanol or ethanol). The biomass remaining from ethanol production is also burnable, and can be used to power the ethanol production instead of using fossil fuels. Many have noted that abnormally high biomass levels are exacerbating risks of forest fires, and have proposed removing much of that biomass to protect forests and communities located near forests. Such woody forest residues could be used to produce energy.

Using wood for energy has some significant drawbacks. Although wood could replace some fossil fuels, it still produces CO_2 (and water vapor and some other by-products) when burned. Wood residues in the forest—from timber harvesting, thinning, or other forestry practices—are widely dispersed; haul distances (and thus costs) may limit the scale of wood energy production facilities. More important, perhaps, is that wood residues are highly variable in size, ranging from tree tops (4 inches or more in diameter) to twigs. Thus, collecting residues is a very labor- intensive activity. The cost to collect and transport forest residues to a wood energy facility can be a major hindrance to using woody forest residues for energy production. In addition, wood energy capacity to use wildfire fuels could exceed the long-term sustainable harvests from the forests, and become redundant or damage future forests. Finally, some wood residue is needed to sustain the productive capacity of forest soils. If too much woody biomass is removed from the forest, soil productivity and future forest growth could decline.

Leakage

Changes in land use practices to sequester carbon in the United States can also have more subtle impacts on carbon storage globally. Practices to store carbon by reducing or delaying timber harvested domestically can have an effect commonly called *leakage*—by shifting land uses geographically (e.g., more tropical forest harvests to offset less temperate forest harvests) or by shifting demand to other products that require more carbon to produce (e.g., steel or aluminum studs to replace wood studs in homebuilding).

Land Use Leakage

The primary concern, expressed in numerous articles, is that forest protection and logging restrictions in the United States will lead to more timber harvesting and associated environmental damage elsewhere—especially in tropical forests—to satisfy U.S. demand for wood products.[50] This leakage is undesirable, it is argued, because U.S. forest management protects the environment more than comparable activities in other countries. These critics assert that U.S. federal and state environmental laws are stricter and more rigorously enforced than other nations' environmental laws for protecting air and water quality, maintaining animal habitats, and preserving rare plants and animals. Thus, they argue, more domestic forest harvesting leads to less forest destruction and greater soil protection, both of which would lead to less carbon release following timber harvesting. In addition, most states (and federal law for federal lands) require reforestation following timber harvests (except for planned conversions to other uses), so deforestation is less common in the United States.

Other reasons are also given for why leakage is undesirable.[51] One is that harvesting in tropical forests typically results in more waste. Tropical forest harvests typically focus on the most valuable species, leaving most of the trees, many of which are damaged; temperate forests have less diversity of plant species, which can lead to greater efficiency in biomass utilization, and thus less biomass waste to return carbon to the atmosphere. Also, temperate forests have less carbon per acre, on average, to release when timber is harvested than do tropical forests. In addition, others note, "[t]imber extraction is often the first step towards opening up the tropical forest and clearing the land for agricultural production. What is more, in many developing countries, property law establishes deforestation as a prerequisite of formal claim over the land for those settling in forested areas."[52] Tropical timber harvests can thus lead to permanent deforestation.

These rationales are supported by substantial anecdotal evidence, but others counter that such assertions have limited empirical foundation. Concerns generally focus on the potential impacts on tropical forests, but the preferred timber species for U.S. consumption are softwoods, not the hardwoods found in the tropics. It is not clear how much tropical timber would substitute for domestic softwood timber.[53] Imports from tree plantations in Chile and New Zealand and from virgin forests in eastern Russia seem more likely, but the environmental effects probably differ substantially from the presumed impacts of leakage to tropical forests.[54] One economist also has pointed out the difficulties in making international comparisons of environmental choices, because of the monumental biological and social differences between the United States and other countries, especially those with tropical forests.[55]

Product Demand Leakage

Another concern often noted is that domestic forest protection to sequester carbon could shift demand to substitute products whose production requires more energy, and thus releases more carbon.[56] Most lumber and plywood are used in construction—new residences, home remodeling, and non-residential buildings.[57] Substitute products include steel and aluminum for studs, joists, and sheathing, and concrete and masonry for walls and flooring. The recent reports expressing concerns about product demand leakage usually cite studies from the 1970s and early 1980s comparing the energy used to produce wood products and their substitutes—notably, studies by the Committee on Renewable Resources for Industrial Materials (CORRIM) of the National Academy of Sciences and by the Office of Technology Assessment of the U.S. Congress.[58] While energy use is not a perfect predictor of carbon release and production technologies have undoubtedly changed substantially in the intervening decades, these studies indicate that wood production requires only about a quarter of the energy needed for concrete and masonry production, and less than 5% of the energy needed for steel or aluminum produced for residential construction. A newer report, depicting average CO_2 emissions, showed somewhat different results, with wood framing, covering, and windows releasing about 30% as much CO_2 as steel and aluminum and about 10% as much CO_2 as concrete and masonry.[59]

Other economists point out that supply substitution is not the only feasible response to changing domestic timber supplies—that demand could be influenced through price changes (although timber is highly price-inelastic), through development of less energy-intensive/carbon-producing non-wood substitutes, and through government policy (e.g., by altering locally-established building codes).[60]

Federal Government Programs

Various federal programs could be used to encourage forestry practices to increase carbon sequestration. One approach is to implement more carbon-sequestering forestry practices on federal lands. Another is to provide technical and financial assistance for forest management practices to private landowners. A third is "tax expenditures" (tax incentives) to encourage carbon-sequestering forestry practices by private landowners.

Federal Forests[61]

The federal government owns about 650 million acres of land in the United States—about 29% of the total U.S. land area.[62] Although this includes hundreds of buildings around the country, the vast majority (more than 99%) is considered rural—national parks and monuments, national wildlife refuges, national forests, public lands, military bases, etc. The majority of the federal lands are in Alaska (3 9%) and the 11 coterminous western states (another 54%).

The Forest Service estimates that the federal government administers 246.4 million acres of forest land—defined as lands available for timber harvesting and capable of growing at least 20 cubic feet of industrial wood per acre annually.[63] Additional federal lands undoubtedly meet the growth standard, but are not included because timber harvesting is precluded by statute (e.g., national parks and wilderness areas) or by administrative decision (e.g., national monuments and inventoried roadless areas). Other federal lands have trees and accumulate wood fiber, but not rapidly enough to meet the growth standard (e.g., Bureau of Land Management lands in central Alaska).

Clearly, forestry practices can and do occur on federal forest lands, affecting both the sequestering and the releasing of carbon, as discussed above. In addition, forestry practices to sequester additional carbon (e.g., planting trees) can occur on sites that would not generally be considered "working" forests (e.g., national parks and military bases). Finally, much of the debate over the carbon consequences of timber harvesting—whether replacing old-growth forests with new, vigorous stands yields new release or net storage of carbon—has focused on federal forests, because most of the remaining old-growth forests occur on federal lands.

Federal Assistance for State and Private Forestry

The federal government has numerous programs that provide technical and financial assistance for forest management of nonfederal (mostly state and private) forest lands, though none explicitly includes carbon sequestration as a purpose.[64] Many programs provide assistance for forestry practices, especially planting trees and improving tree growth. Some provide technical assistance, others provide financial assistance (usually a share of the activity's cost), and a few provide both. Some of the programs are foresty-specific (and administered by the Forest Service), while others are agricultural conservation programs that include forestry activities as acceptable treatments.[65] Most are at least coordinated with, if not funded and operated through, state forestry agencies. It should be recognized, however, that many of these programs are restricted to moderately or highly productive private forest lands.

The federal government also has programs to assist in protecting non-federal forests. One focuses on identifying and controlling insect and disease infestations, run by the Forest Service, in cooperation with the states. Another emphasizes preventing and controlling catastrophic wildfires, through assistance (money, equipment, and technical help) to states and to rural volunteer fire departments. Forest protection programs, such as the Forest Legacy, provide federal funding for the federal government or for states to purchase title or easements on forestlands that might be converted to other uses (e.g., agriculture or housing).

Finally, the Forest Service is also authorized to provide technical forestry assistance to other countries. This international forestry program includes information about carbon sequestration as well as about forestry practices that yield other economic and social benefits.

Federal Tax Expenditures

"Tax expenditures" are specific tax incentives established to encourage or allow certain activities. Three federal tax expenditures apply to forestry practices. One is special treatment of reforestation expenses: a tax credit and accelerated amortization of annual reforestation expenses.[66] Private landowners are allowed to take a tax credit for 10% of their annual reforestation costs, up to $1,000 credit per year, and to amortize their reforestation costs, up to $10,000 per year, over an 84-month period. With reforestation costs averaging about $250 per acre (generally higher in the West and lower in the South), this provision would apply to about 40 acres of reforestation annually—not significant to major corporate forestland owners (who may own more than a million acres of forestland), but substantial for many of the non-industrial private landowners who own nearly half of the forestland in the United States.

The second special provision allows the expensing of multi-period timber growing costs; that is, annual management expenses can be deducted from other current income, rather than capitalized and deducted from timber income from the managed stands. As with the accelerated amortization for reforestation, this reduces the current income tax liability for the private landowners—a significant benefit for all private forestland owners. It also simplifies their bookkeeping for tax purposes—a significant benefit for the non-industrial landowners with modest forestland holdings and only occasional timber income.

Finally, timber income is allowed to be treated as a capital gain, if the private landowner has owned the trees for more than a year and does not retain an economic interest in the trees after the sale. (This latter provision essentially precludes certain timber sale practices commonly used by the federal government.) Currently, this provision benefits private forestland owners, because capital gains tax rates are significantly lower than regular income tax rates.

These three tax expenditures affect carbon-sequestering forestry practices in several ways. The reforestation tax credit and accelerated depreciation and the expensing of multi-period timber growing costs effectively reduce the landowners' costs of these forestry practices, which can induce additional forestry practices and thus may add to carbon sequestration. Capital gains treatment of timber income likely encourages timber harvesting, but as discussed above, the carbon consequences of timber harvesting are disputed.

Federal Programs Affecting Land Use

Many federal programs, in addition to the programs discussed above, can affect the rate of deforestation and afforestation in the United States. Development on forest land, often called "urban sprawl," is affected by a wide variety of federal programs, such as transportation and other infrastructure assistance programs and various tax expenditures. The U.S. General Accounting Office (now the Government Accountability Office) examined these programs, finding that federal impacts are pervasive, but that changing federal policies might not have major impacts.[67] The federal programs directly targeted on reducing the development of forest land are the Forest Legacy and the Community Forest and Open Conservation Program, both of which fund government or nonprofit acquisition of title or easements for forestland threatened with development (i.e., conversion to non-forest uses).

Numerous agricultural conservation programs affect the conversion of forests to and from pasture and cropland.[68] Most notable is the USDA's Conservation Reserve Program (CRP).[69] The CRP pays farmers who own environmentally sensitive and highly erodible cropland, under multi-year contracts, to protect those lands; planting trees is a common method for protecting these croplands, and the CRP is the largest federal tree planting program that has ever existed, even though its primary purpose is to protect soils. Other USDA programs that might include funding for afforestation include the Environmental Quality Enhancement Program (EQIP), the Farmland Protection Program, and the Wildlife Habitat Incentives Program.

Accounting for Forest Carbon Sequestration

Section 1605(b) of the Energy Policy Act of 1992 (P.L. 102-486) established a purely voluntary system to collect information from entities that emit greenhouse gases, including information on how they are reducing emissions or sequestering carbon by "any measures, including fuel switching, forest management practices, tree planting," and more. The Energy Information Agency established guidelines for data reporting and self-certification. While most of the emissions reductions reported in the § 1605(b) program are related to energy efficiency or conservation, some are forestry projects—in other countries (known as "joint implementation") as well as within the United States. Relatively few domestic forestry projects have been reported under the § 1605(b) program. A reason to report under the program has been the expectation (or hope) of receiving retroactive credit if a carbon credit system were ever established. Whether such credit would be given and whether reported data are deemed sufficiently reliable remain open, and possibly contentious, questions.

CONCLUSION

Forests store substantial amounts of carbon. The amount stored, however, changes over time as forests grow and die. Land use changes and forestry practices alter the level and rates of carbon storage, while "leakage" (shifting production) may offset some of the increases in forest carbon sequestration. Whether and how to account for this carbon sequestration in policies and programs to mitigate climate change has been controversial. Under the 1997 Kyoto Protocol of the 1992 United Nations Framework Convention on Climate Change (UNFCCC), many developed nations agreed to specified reductions in their emissions of greenhouse gases, especially carbon dioxide. The Protocol allows some carbon sequestration as a way of meeting the specified reductions. The Bush Administration rejected the Protocol and withdrew from the continuing activities under the Protocol. Nonetheless, accounting for the carbon absorbed by forests and soils (and how much credit is due) continues to be discussed internationally; and the U.S. Congress has held hearings on forest carbon sequestration.

The role of forestry and land use in mitigating climate change has been controversial. Forests are enormously variable, with a broad array of plant species (both trees and understory vegetation) and substantial differences in the diversity of plant (and animal) species they contain. The myriad permutations of forest plants and soils present formidable obstacles for estimating existing carbon stocks and the carbon sequestration and release that result from forestry activities. The carbon consequences of timber harvesting have been particularly controversial.

Because of the scientific uncertainties, as well as differences in the types and extent of forests among nations, reaching agreement on ways to account for carbon sequestration in forests has been difficult. Some argue for a broadly inclusive accounting, others for a more conservative approach. "Land-based" or "activity-based" models are generally proposed for estimating changes in carbon storage. However, the ambiguous language and terminology used by proponents contribute to the inherent difficulties of measuring baseline carbon stocks, land uses, the carbon impacts of various activities, and "leakage" (shifting land or product uses). Furthermore, the enormous diversity of forest types and widespread disputes over the carbon consequences of various practices (which result at least partly from the diversity of forests) make it difficult to generalize about the opportunities to mitigate global climate change through forest carbon sequestration. It is likely that research to reduce some of these ambiguities and uncertainties will be an ongoing element in the efforts of nations to deal with carbon sequestration—and with concerns about climate change.

End Notes

[1] This report does not address underlying questions of whether global warming is occurring or of the possible human role. See CRS Report RL33849, *Climate Change: Science and Policy Implications*, by Jane A. Leggett.

[2] This report does not address the impacts of climate change on forests, although this is also an important scientific issue. For a discussion of these impacts, see the series of articles in "A Special Section on Climate Change and Forest Ecosystems," *Bioscience*, v. 51, no. 9 (Sept. 2001): 720-779.

[3] See CRS Report RL33826, *Climate Change: The Kyoto Protocol, Bali "Action Plan," and International Actions*, by Jane A. Leggett.

[4] See CRS Report RS22806, *The Bali Agreements and Forests*, by Ross W. Gorte and Pervaze A. Sheikh.

[5] See CRS Report RL3 1286, *Debt-for-Nature Initiatives and the Tropical Forest Conservation Act: Status and Implementation*, by Pervaze A. Sheikh.

[6] Fire is a self-sustaining chemical process that breaks organic chemicals down into minerals, water, and CO_2.

[7] Roots less than 2 millimeters in diameter are generally considered to be part of the soil, not part of the plant that grew them.

[8] Guoyi Zhou et al., "Old-Growth Forests Can Accumulate Carbon in Soils," *Science*, v. 314 (Dec. 1, 2006): 1417. Hereafter referred to as Zhou et al., "Soil Carbon in Old-Growth Forests."

[9] This is one reason why these species are preferred for timber plantations—a greater proportion of total biomass production goes into commercial wood, and is not "wasted" on noncommercial biomass.

[10] One study—Paul Schroeder, "Can Intensive Management Increase Carbon Storage in Forests?" *Environmental Management*, v. 15, no. 4 (1991): 475-481, hereafter referred to as Schroeder, "Intensive Management for Carbon Storage"—assumed a "biomass expansion factor" of 1.6; that is, it assumed that total biomass was 60% greater than (1.6 times) the commercial wood biomass. Another study—Jack K. Winjum, Sandra Brown, and Bernhard Schlamadinger, "Forest Harvests and Wood Products: Sources and Sinks of Atmospheric Carbon Dioxide," *Forest Science*, v. 44, no. 2 (1998): 272-284—assumed biomass expansion factors of 1.3 for conifer forests and 2.0 for non- conifer forests. A third study—Robert J. Moulton and Kenneth R. Richards, *Costs of Sequestering Carbon Through Tree Planting and Forest Management in the United States*, USDA Forest Service, Gen. Tech. Rept. WO-58, Washington, DC, Dec. 1990—used biomass expansion factors ranging from less than 2.0 to more than 8.0 just for different forests within the United States.

[11] A "biome" is defined as a "[r]egional land-based ecosystem type ... characterized by consistent plant forms and ... found over a large climatic area." Examples include tropical rainforests,

tundra, temperate grasslands, deserts, etc. From *The Dictionary of Ecology and Environmental Science,* Henry W. Art, ed., Henry Holt & Co., New York, NY, 1993, p. 65.

12 Data on carbon stocks presented in this section are CRS calculations from data in Intergovernmental Panel on Climate Change, "Table 1: Global carbon stocks in vegetation and carbon pools down to a depth of 1 m [meter]," *Summary for Policymakers: Land Use, Land-Use Change, and Forestry. A Special Report of the Intergovernmental Panel on Climate Change*, at *http://www.ipcc.ch/pub/srlulucf-e.pdf.* Hereafter referred to as IPCC, *Special Report.*

13 "Hardwoods" is a term commonly used for trees that are angiosperms—flowering plants—because the dominant flowering tree species of temperate climates (oaks and maples) are harder (more dense) than the major "softwood" species (pines, firs, and spruces), trees of the order Coniferales (conifers). However, some "hardwood" species (e.g., aspen and poplar) are much softer (less dense) than many "softwoods." In temperate areas, most hardwoods are also deciduous (losing all their leaves annually), while most conifers are evergreen (retaining their needles for more than one year), leading to common use of "deciduous" and "evergreen" as synonyms for angiosperms and conifers. However, certain conifers (notably larches) are deciduous, while many hardwoods in subtropical climates and most in tropical climates are evergreen. For this report, "hardwood" is used to indicate angiosperms, while "softwood" (or conifer) is used for coniferous species.

14 Soil carbon levels in wetlands are nearly double the level in boreal forests, because the frequent standing water prevents aerobic decomposition, and anaerobic (without oxygen) decomposition processes are much slower than aerobic processes.

15 Richard Birdsey, Ralph Alig, and Darius Adams, "Chapter 8. Mitigation Activities in the Forest Sector to Reduce Emissions and Enhance Sinks of Greenhouse Gases," *The Impact of Climate Change on America's Forests: A Technical Document Supporting the 2000 USDA Forest Service RPA Assessment*, USDA Forest Service, Rocky Mountain Research Station, Gen. Tech. Rept. RMRS-GTR-59, Fort Collins, CO, 2000, p. 116. Hereafter referred to as Birdsey et al., "Forest Mitigation."

16 IPCC, *Special Report*, "4. Carbon Accounting."

17 IPCC, *Special Report*, paragraph 31.

18 See also CRS Report RS22964, *Measuring and Monitoring Carbon in the Agricultural and Forestry Sectors*, by Ross W. Gorte and Renée Johnson.

19 Birdsey et al., "Forest Mitigation," p. 112.

20 CRS calculations from data in IPCC, *Special Report*, table 1, p. 4.

21 CRS calculations from data in IPCC, *Special Report*, table 1, p. 4.

22 USDA Natural Resources Conservation Service and Iowa State University Statistical Laboratory, *Summary Report: 1997 National Resources Inventory* (Washington, DC: Dec. 1999), p. 35. More recent inventories do not contain comparable data on land conversions.

23 U.S. House, Committee on Resources, Subcommittee on Forest and Forest Health, *H.Con.Res. 151, Expressing the Sense of the Congress that the United States Should Manage Its Public Domain National Forests to Maximize the Reduction of Carbon Dioxide in the Atmosphere Among Many Other Objectives and that the United States Should Serve as an Example and as a World Leader in Actively Managing Its Public Domain National Forests in a Manner that Substantially Reduces the Amount of Carbon Added to the Atmosphere*, Serial No. 105-6 1, GPO, Washington, DC, 1998.

24 See Mark E. Harmon, William K. Ferrell, and Jerry F. Franklin, "Effects on Carbon Storage of Conversion of Old- Growth Forests to Young Forests," *Science*, v. 247 (Feb. 9, 1990): 699-702, hereafter referred to as Harmon, et al., "Conversion of Old Growth Forests"; Robert C. Musselman and Douglas G. Fox, "A Review of the Role of Temperate Forests in the Global CO2 Balance," *Journal of the Air and Waste Management Association*, v. 41, no. 6 (June 1991): 798-807, hereafter referred to as Musselman and Fox, "Role of Temperate Forests"; and Andrew J. Plantinga, Thomas Mauldin, and Douglas J. Miller, "An Econometric

Analysis of the Costs of Sequestering Carbon in Forests," *American Journal of Agricultural Economics*, v. 81, no. 4 (Nov. 1, 1999).

[25] Roderick C. Dewar and Melvin G. R. Cannell, "Carbon Sequestration in the Trees, Products and Soils of Forest Plantations: An Analysis Using UK Examples," *Tree Physiology*, v. 11 (1992): 49-71. Summary at *http://heronpublishing.com/tree/summaries/volume11/a11-49.html.*

[26] R. A. Birdsey and G. M. Lewis, *Carbon in U.S. Forests and Wood Products, 198 7-1997: State-by-State Estimates*, USDA Forest Service, Northeast Research Station and U.S. Environmental Protection Agency, Gen. Tech. Rept. NE- 310, Newton Square, PA, 2002.

[27] R. A. Houghton, J. L. Hackler, and K. T. Lawrence, "The U.S. Carbon Budget: Contributions from Land-Use Change," *Science*, v. 285 (July 23, 1999): 577.

[28] Musselman and Fox, "Role of Temperate Forests," p. 803.

[29] Harmon, et al., "Conversion of Old-Growth Forests."

[30] For background information, see CRS Report RL3 0755, *Forest Fire/Wildfire Protection*, by Ross W. Gorte.

[31] Bill Leenhouts, "Assessment of Biomass Burning in the Conterminous United States," *Conservation Biology* [online], v. 2, no. 1 (1998); available at *http://www.consecol.org/vol2/iss1/art1.*

[32] Philip N. Omi and Erik J. Martinson, *Effects of Fuels Treatment on Wildfire Severity: Final Report*, submitted to the Joint Fire Science Program Governing Board, Colorado State Univ., Western Forest Fire Research Center, Ft. Collins, CO, March 25, 2002, p. I.

[33] Henry Carey and Martha Schumann, *Modifying WildFire Behavior—The Effectiveness of Fuel Treatments: The Status of Our Knowledge*, National Community Forestry Center, Southwest Region Working Paper 2, Santa Fe, NM, April 2003.

[34] J. T. Randerson et al., "The Impact of Boreal Forest Fire on Climate Warming," *Science*, v. 314 (Nov. 17, 2006): 1130-1132.

[35] See David M. Smith et al., *The Practice of Silviculture: Applied Forest Ecology*, John Wiley & Sons, Inc., New York, NY, 1997.

[36] Schroeder, "Intensive Management for Carbon Storage."

[37] Schroeder, "Intensive Management for Carbon Storage."

[38] See, for example, E. H. DeLucia, D. J. Moore, and R. J. Norby, "Contrasting Responses of Forest Ecosystems to Rising Atmospheric CO2: Implications for the Global C [Carbon] Cycle," *Global Biogeochemical Cycles*, v. 19 (July 19, 2005).

[39] Adrien C. Finzi et al., "Progressive Nitrogen Limitation of Ecosystem Processes Under Elevated CO_2 in a Warm- Temperature Forest," *Ecology*, v. 87, no. 1 (Jan. 2006): 15-25.

[40] Prescribed burning is typically only used in ecosystems adapted to relatively frequent surface fires (5- to 35-year intervals between fires), which burn the grasses, needles, twigs, and undergrowth while allowing most of the trees to survive. Such *frequent-surface-fire* ecosystems occur primarily in temperate zones, especially where lightning occurs with little or no rainfall. However, not all temperate forest ecosystems have adapted to relatively frequent surface fires. Some, such as lodgepole pine, jack pine, and aspen, are adapted to regenerate naturally following periodic "crown" fires, where many or most of the trees are killed. In such ecosystems, prescribed burning would poorly simulate natural ecological patterns.

[41] Harmon et al., "Conversion of Old-Growth Forests."

[42] Kenneth E. Skog and Geraldine A. Nicholson, "Carbon Sequestration in Wood and Paper Products," in *The Impact of Climate Change on America's Forests: A Technical Document Supporting the 2000 USDA Forest Service RPA Assessment*, USDA Forest Service, Rocky Mountain Research Station, Gen. Tech. Rept. RMRS-GTR-59, Fort Collins, CO, 2000, pp. 79-88. Hereafter referred to as Skog and Nicholson, "Carbon Sequestration in Products."

[43] CRS could find no source that identifies the proportion of carbon from wood pulp converted into paper products.

[44] Skog and Nicholson, "Carbon Sequestration in Products." However, use of biomass waste to generate heat and/or electricity can substitute for fossil fuel use, thereby avoiding other CO_2 emissions.

[45] See John Perez-Garcia, Chadwick D. Oliver, and Bruce R. Lippke, "How Forests Can Help Reduce Carbon Dioxide Emissions to the Atmosphere," (July 7, 1997) in U.S. House, Committee on Resources, Subcommittee on Forest and Forest Health, *Hearing on H.Con.Res. 151*, Serial No. 105-61, GPO, Washington, DC, 1998, pp. 46-68. Hereafter cited as Perez-Garcia et al., "How Forests Reduce Carbon Dioxide Emissions."

[46] Harmon, et al., "Conversion of Old-Growth Forests"; and Peter M. Vitousek, "Can Planted Forests Counteract Increasing Atmospheric Carbon Dioxide?" *Journal of Environmental Quality*, v. 20 (April-June 1991): 348-354.

[47] Byung Bae Park, *Effects of Silvicultural Treatments on Carbon Storage in Northern Hardwood Forests*, available at *http://www.esf.edu/resorg/rooseveltwildlife/Research/Silv/Silv.htm.*

[48] Zhou, et al., "Soil Carbon in Old-Growth Forests."

[49] U.S. Congress, Office of Technology Assessment, *Wood Use: U.S. Competitiveness and Technology, Volume II— Technical Report*, OTA-M-224, Washington, DC, Nov. 1984, p. 18. Hereafter cited as OTA, *Wood Use*.

[50] See, e.g., Con H. Schallau and Alberto Goetzl, "Effects of Constraining US Timber Supplies: Repercussions Are National and Global," *Journal of Forestry*, v. 90, no. 7 (July 1992): 22-27; Roger A. Sedjo et al., *Global Forest Products Trade: The Consequences of Domestic Forest Land-Use Policy*, Resources for the Future, Discussion Paper 94-13, Washington, DC, Feb. 1994; and John M. Perez-Garcia, "Global Economic and Land Use Consequences of North American Timberland Withdrawals," *Journal of Forestry*, v. 93, no. 7 (July 1995): 34-38.

[51] See Ed Barbier, "The Environmental Effects of Trade in the Forestry Sector," *The Environmental Effects of Trade*, Organisation for Economic Co-operation and Development [OECD], Paris, France, 1994, pp. 55-101.

[52] Ibid., p. 67.

[53] Research on Japanese timber consumption (which is more diverse than U.S. timber consumption) showed limited substitution between well-known and lesser-known species from tropical forests. See Jeffrey R. Vincent, Alamgir K. Gandapur, and David J. Brooks, "Species Substitution and Tropical Log Imports by Japan," *Forest Science*, v. 36, no. 3 (Sept. 1990): 657-664. However, as tropical log supplies have dwindled, substitutability among tropical species might have risen in Japan.

[54] See David J. Brooks, "Federal Timber Supply Reductions in the Pacific Northwest: International Environmental Effects," *Journal of Forestry*, v. 93, no. 7 (July 1995): 29-33.

[55] David J. Brooks, *U.S. Forests in a Global Context*, USDA Forest Service, Rocky Mountain Research Station, Gen. Tech. Rept. RM-228, Fort Collins, CO, July 1993.

[56] See, e.g., Perez-Garcia et al., "How Forests Reduce Carbon Dioxide Emissions," and R. Neil Sampson, "Forest and Wood Products Role in Carbon Sequestration," in U.S. House, Committee on Resources, Subcommittee on Forest and Forest Health, *Hearing on H.Con.Res. 151*, Serial No. 105-61, GPO, Washington, DC, 1998, pp.34-45.

[57] Western Wood Products Association, *2007 Statistical Yearbook of the Western Lumber Industry*, Portland, OR, p. 25.

[58] National Academy of Sciences, Committee on Renewable Resources for Industrial Materials, *Renewable Resources for Industrial Materials*, Washington, DC: 1976, 266 p.; National Academy of Sciences, Committee on Renewable Resources for Industrial Materials, "Wood and Fiber: Special CORRIM Panel II Report," *Journal of the Society of Wood Science and Technology*, v. 8, no. 1 (spring 1976): 1-72; and OTA, *Wood Use*.

[59] Cooperative Research Centre for Greenhouse Accounting (CRC), *Counting Carbon: Wood Products*, at *http://www.greenhouse*. The CRC has ceased operations, but its materials are still being made available by Australian National University.

[60] David J. Brooks, "International Dimensions of U.S. Forestry," *Agriculture Outlook '93 Conference* (Washington, DC: U.S. Dept. of Agriculture, Dec. 2, 1992), pp. 19-33.

[61] For a description of federal lands and the agencies that administer them, see CRS Report R40225, *Federal Land Management Agencies: Background on Land and Resources Management*, coordinated by Ross W. Gorte.

[62] U.S. General Services Administration, *Federal Real Property Profile As of September 30, 2004*, Washington, DC, pp. 18-19.

[63] W. Brad Smith et al., *Forest Resources of the United Stataes, 2002*, USDA Forest Service, North Central Research Station, Gen. Tech. Rept. NC-241, St. Paul, MN, 2004, pp. 31-32, *http://www.ncrs.fs.fed.us/pubs/gtr/gtr_nc241.pdf.*

[64] See CRS Report RL3 1065, *Forestry Assistance Programs*, by Ross W. Gorte.

[65] For information on agricultural conservation programs, see CRS Report RL34557, *Conservation Provisions of the 2008 Farm Bill*, by Tadlock Cowan, Renée Johnson, and Megan Stubbs.

[66] Other reforestation expenses must be capitalized, and carried on the books to be deducted against timber income when the reforested stand is harvested, 20-100 years or more later.

[67] U.S. General Accounting Office (now Government Accountability Office), *Community Development: Extent of Federal Influence on "Urban Sprawl" Is Unclear*, GAO/RCED-99-87, Washington, DC, April 1999.

[68] CRS Report RL34557, *Conservation Provisions of the 2008 Farm Bill*, by Tadlock Cowan, Renée Johnson, and Megan Stubbs.

[69] See CRS Report RS21613, *Conservation Reserve Program: Status and Current Issues*, by Tadlock Cowan.

In: Carbon Considerations
Editors: W. Green and L. Wayman ISBN 978-1-62618-435-0

Chapter 2

BIOCHAR FOR SOIL FERTILITY AND NATURAL CARBON SEQUESTRATION*

Collen E. Rostad and David W. Rutherford

INTRODUCTION

Biochar is charcoal (similar to chars generated by forest fires) that is made for incorporation into soils to increase soil fertility while providing natural carbon sequestration. The incorporation of biochar into soils can preserve and enrich soils and also slow the rate at which climate change is affecting our planet. Studies on biochar, such as those cited in this Fact Sheet, are applicable to both fire science and soil science.

When plant materials—such as wood, bark, corn stalks, switchgrass, and the like, which scientists generically term biomass—are heated in a low- or no-oxygen atmosphere (this process is termed pyrolysis), the resulting product is called biochar. For example, pyrolysis of wood yields a familiar biochar: charcoal, once the staple fuel for backyard cookouts.

Carbon in the form of carbon dioxide (CO_2), which is released into the atmosphere when carbon-based fuels are burned and when biomass decays, is a major contributor to global climate change. For this reason, carbon sequestration (the process of capturing CO_2 before it escapes into the atmosphere) has

* This fact sheet was released January 2011 by the U.S. Department of the Interior, U.S. Geological Survey.

the potential to slow global warming. Increasingly, scientists are looking at biochars as additives that can simultaneously enrich soil and sequester carbon that would otherwise be released into the atmosphere if biomass were simply left aboveground to decay.

Douglas-fir mortality caused by Douglas-fir beetle in the Uncompahgre National Forest near Telluride, Colorado. Photograph taken by Justin Backsen of the U.S. Forest Service.

Plant biomass is primarily a mixture of cellulose and lignin with small amounts of inorganic material called ash. Waste biomass such as beetle-killed trees or agricultural residues can be put to beneficial use by conversion to biochar.

The pyrolysis of plant biomass to generate biochar converts much of the aliphatic carbon (which contains no carbon ring structures) to a very stable form of aromatic carbon (which contains rings made of carbon, like graphite) which is stable in soils for hundreds of years. The term biochar has been adopted to mean char produced from waste biomass for the purpose of land application.

The application of biochar to soil has been proposed as a way to store carbon in the soil for longer periods than if biomass was left to decay. Besides increasing the storage of carbon in soils, the application of biochar to soil can improve soil fertility and increase crop production. Since biochar is different from better-known, activated charcoal, information is needed on the effects of its formation conditions on both its physical and chemical properties before farmers begin to use it for large-scale incorporation into soils.

One of a nation's most important assets is the fertility of its soil. The cultivation of soil causes a steady decrease in the soil's organic content and fertility. This steady decrease in fertility is compensated for by addition of fertilizers, but the decrease in organic content is a more permanent problem. Incorporation of biochar can slow or reverse this trend.

Pine shavings, before and after charring.

As a means of predicting biochar characteristics, the U.S. Geological Survey (USGS) made biochar by charring a variety of feedstocks including purified plant components (such as cellulose and lignin) and plant materials (such as wood, bark, and switchgrass) under a nitrogen atmosphere at temperatures ranging from 200°C to 500°C (about 400°F to 900°F) for times ranging from 1 hr (hour) to 168 hr. Changes in composition were examined by mass loss, elemental composition (carbon, hydrogen, and oxygen), acid-functional-group content, and infrared and ^{13}C Nuclear Magnetic Resonance (NMR) spectroscopy (Rutherford and others, 2008). Changes in structure were examined by porosity and surface area (Rutherford and others, 2004).

The USGS has obtained bulk quantities of biochar reference material from two gasification processes which produce biochar as a byproduct. The reference materials were homogenized and split to ensure uniformity. The uniformity of these materials was verified by sub-sampling and subsequent chemical and physical analyses. These two reference chars provide a consistent material to the scientific research community for comparison of analytical techniques between laboratories or as a basis for comparison of characteristics of one type of char to another.

The two main families of carbon compounds are aliphatic and aromatic. Comparison of mass loss, combined with ^{13}C NMR spectroscopy, showed that the initial rapid loss of material during pyrolysis is due to the loss of aliphatic components, which are either vaporized or converted to aromatic carbon early in the charring process. Aromatic carbon is a more stable form than aliphatic carbon.

The biochars were also analyzed for acid functional group content. Acid functional groups are related to cation-exchange capacity, which is important to soil fertility, as it assists in the retention of nutrients (nitrogen and phosphorus) and cations (like potassium) in soil. In the USGS experiments, the highest concentrations of acid functional groups developed at low charring temperatures (250°C) and long charring times (greater than 72 hr). Cellulose appears to be the primary source of the acid functional groups in the biochar.

Surface area and porosity of the biochar plays a significant role in soil fertility. In contrast to the optimum conditions for the formation of the acid functional groups, more intense charring conditions (higher temperatures and longer charring times) were required for the formation of porosity and surface area in the biochar.

The effect of formation conditions and feedstock on physical and chemical characteristics of the resultant char will be important if biochar is to be successfully incorporated into agricultural soils. Using this information provided by USGS, formation conditions and feedstocks may be tailored to optimize desirable characteristics (such as acid functional group content) and to minimize less-desirable characteristics (such as degradability or mass loss). The results generated by USGS provide insight into optimum formation conditions for chars with the best properties to simultaneously enhance soil fertility and provide longterm carbon sequestration.

References

Rutherford, D.W., Wershaw, R.L., and Reeves, J.B., III, 2008, Development of acid functional groups and lactones during the thermal degradation of wood and wood components: U.S. Geological Survey Scientific Investigations Report 2007–5013, 43 p. (Also available at http://pubs.usgs.gov/sir/2007/5013)

Rutherford, D.W., Wershaw, R.L., and Cox, L.G., 2004, Changes in composition and porosity occurring during the thermal degradation of wood and wood components: U.S. Geological Survey Scientific Investigations Report 2004–5292, 78 p. (Also online at http://pubs.usgs.gov/sir/2004/5292/.)

In: Carbon Considerations
Editors: W. Green and L. Wayman
ISBN 978-1-62618-435-0

Chapter 3

BIOCHAR: EXAMINATION OF AN EMERGING CONCEPT TO SEQUESTER CARBON*

Kelsi Bracmort

SUMMARY

Biochar is a charcoal produced under high temperatures using crop residues, animal manure, or any type of organic waste material. Depending on the feedstock, biochar may look similar to potting soil or to a charred substance. The combined production and use of biochar is considered a carbon-negative process, meaning that it removes carbon from the atmosphere.

Biochar has multiple potential environmental benefits, foremost the potential to sequester carbon in the soil for hundreds to thousands of years at an estimate. Studies suggest that crop yields can increase as a result of applying biochar as a soil amendment. Some contend that biochar has value as an immediate climate change mitigation strategy. Scientific experiments suggest that greenhouse gas emissions are reduced significantly with biochar application to crop fields.

Obstacles that may stall rapid adoption of biochar production systems include technology costs, system operation and maintenance, feedstock availability, and biochar handling. Biochar research and

* This is an edited, reformatted and augmented version of a Congressional Research Service publication, CRS Report for Congress R40186, from www.crs.gov, prepared for Members and Committees of Congress, dated January 11, 2011.

development is in its infancy. Nevertheless, interest in biochar as a multifaceted solution to agricultural and natural resource issues is growing at a rapid pace both nationally and internationally.

Past Congresses have proposed numerous climate change bills, many of which do not directly address mitigation and adaptation technologies at developmental stages, such as biochar. However, biochar may equip agricultural and forestry producers with numerous revenue-generating products: carbon offsets, soil amendments, and energy.

This report briefly describes biochar, some of its potential advantages and disadvantages, legislative support, and research and development activities underway in the United States.

INTRODUCTION

Biochar—a charcoal produced under high temperatures using crop residues, animal manure, or other organic material—has the potential to offer multiple environmental benefits. Some contend that biochar can meet pressing environmental demands by sequestering large amounts of carbon in soil. It is of interest to those seeking to sell or purchase carbon offsets, increase soil conservation efforts, improve crop yield, and produce renewable energy. However, little is known about how biochar production systems could successfully be implemented and what the effect would be on long-term operations in the U.S. agriculture and forestry sectors. Some contend that it will be a considerable amount of time before this technology reaches its full potential. Studies underway at federal government research institutions and in academia are focused on ensuring that biochar production systems are a practical and reliable technology for producers to adopt.

BIOCHAR

Biochar is a soil supplement that sequesters carbon in the soil and thus may help to mitigate global climate change.[1] It has the potential to curtail greenhouse gas emissions and other environmental hazards in the near term and to benefit agricultural producers as a soil amendment and source of renewable energy. Thus far, biochar use in the United States has been confined to small-scale applications reflective of the limited but growing number of researchers in this area over the last few years.

Biochar is similar in appearance to potting soil or to a charred substance, depending on the feedstock (*Figure 1*). Modern biochar production is based on an ancient Amazon technique for creating a nutrient-rich soil, *terra preta*. As a charcoal containing high levels of organic matter, biochar is formed from plant and crop residues or animal manure under pyrolysis conditions. Pyrolysis is the chemical breakdown of a substance under extremely high temperatures in the absence of oxygen.

Source: Biochar Engineering (BEC Inc.; http://www.biocharengineering.com).

Notes: Wood chips on the right and in barrel were processed through a biochar production system.

Figure 1. Biochar.

The quantity and quality of biochar production depends on the feedstock, pyrolysis temperature, and pyrolysis processing time. A "fast" pyrolysis (~500°C) produces biochar in a matter of seconds, while a "slow" pyrolysis produces considerably more biochar but in a matter of hours.[2]

Biochar production via pyrolysis is considered a carbon-negative process because the biochar sequesters carbon while simultaneously enhancing the fertility of the soil on which the feedstock used to produce the bioenergy grows (*Figure 2*). The biochar production system is operated using energy produced by the system. The three main outputs of a biochar production system are syngas, bio-oil, and biochar (*Figure 3*).[3]

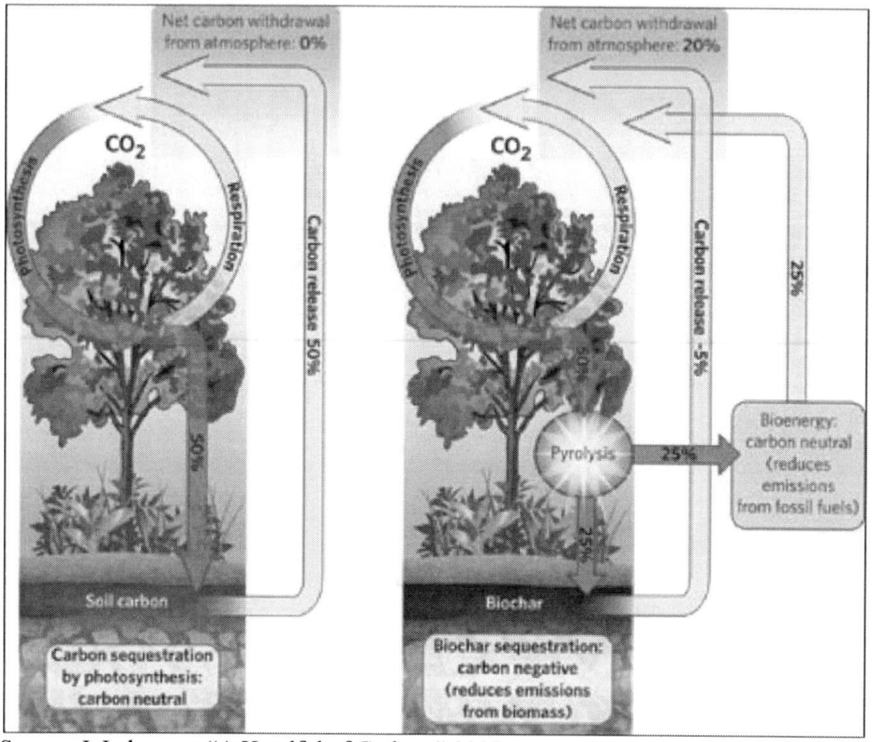

Source: J. Lehmann, "A Handful of Carbon," 2007. Nature 447, pp. 143-144.

Figure 2. Carbon Cycle for Soil Carbon and Biochar.

Potential Advantages

Whether used as a soil amendment[4] or burned as an energy source (e.g., for cooking and heating), biochar provides numerous potential environmental benefits, some of which are not quantifiable. The three primary potential benefits are carbon sequestration, greenhouse gas emission reduction, and soil fertility.

Carbon Sequestration

Carbon sequestration is the capture and storage of carbon to prevent it from being released to the atmosphere. Studies suggest that biochar sequesters approximately 50% of the carbon available within the biomass feedstock being pyrolyzed, depending upon the feedstock type.[5] The remaining carbon is released during pyrolysis and may be captured for energy production. Large amounts of carbon may be sequestered in the soil for long time periods (hundreds to thousands of years at an estimate),[6] but precise estimates of carbon amounts sequestered as a result of biochar application are scarce. One scientist suggests that a 250-hectare farm could sequester 1,900 tons of CO2 a year.[7]

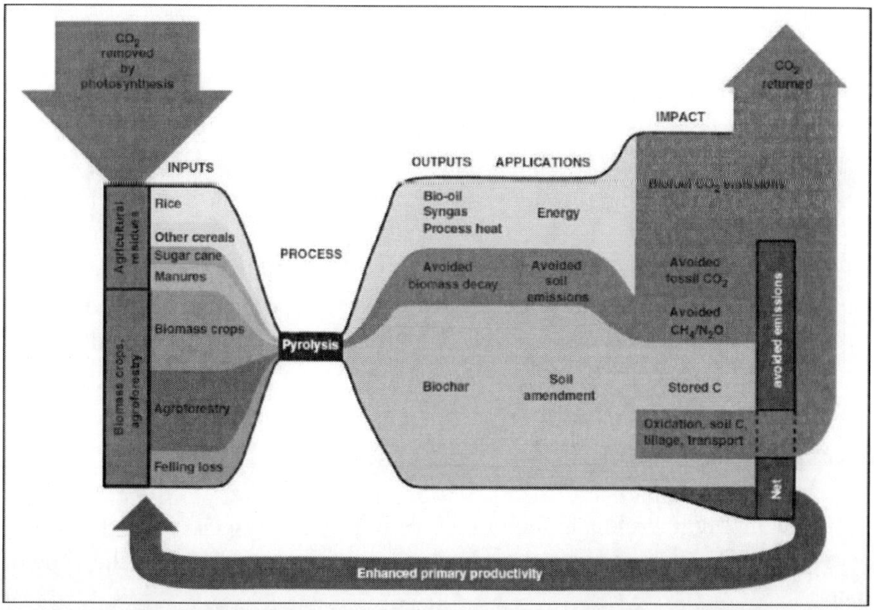

Source: Dominic Woolf, James E. Antonette, and F. Alayne Street-Perrott, et al., "Sustainable Biochar to Mitigate Global Climate Change," Nature Communications, August 10, 2010.

Notes: Within each category (inputs, process, outputs, applications and impacts on global climate), the relative proportions of the components are approximated by the height and width of the colored fields.

Figure 3. Biochar Inputs, Process, Outputs,Applications, and Global Climate Impact.

Greenhouse Gas Emission Reduction

Primary greenhouse gases associated with the agriculture sector are nitrous oxide (N_2O) and methane (CH_4). Cropland soils and grazing lands are an agricultural source of nitrous oxide emissions.[8] Livestock manure management and enteric fermentation are leading agricultural sources of methane emissions.[9] When applied to the soil, biochar can lower greenhouse gas emissions by substantially reducing the release of nitrous oxide.[10] One report showed a 40% reduction in emissions of this greenhouse gas, which is approximately 310 times stronger than carbon dioxide in terms of global warming potential.[11] Laboratory studies suggest that nitrous oxide emission reductions from biochar-treated soil are dependent on soil moisture and soil aeration.[12] Greenhouse gas emission reductions may be 12%-84% greater if biochar is land-applied instead of combusted for energy purposes.[13]

Soil Fertility

Biochar retains nutrients for plant uptake and soil fertility. The infiltration of harmful quantities of nutrients and pesticides into groundwater and soil erosion runoff into surface waters can be limited with the use of biochar.[14] If used for soil fertility, biochar may have a positive impact on those in developing countries. Impoverished tropical and subtropical locales with abundant plant material feedstock, inexpensive cooking fuel needs, and agricultural soil replenishment needs could see an increase in crop yields.[15]

Potential Disadvantages

Recognizing that biochar technology is in its early stages of development, there are many concerns about the applicability of the technology in the United States.

Three issues are feedstock availability, biochar handling, and biochar system deployment. Successful implementation of biochar technology depends on the ability of the agricultural community to afford and operate a system that is complementary to current farming and forestry practices.

Feedstock Availability

The availability of a plentiful feed supply for biochar production is an area for further study. To date, feedstock for biochar has consisted mostly of plant and crop residues, a primary domain of the agricultural community. There may be a role for the forestry community to be involved, as woody biomass is deemed a cost-effective, readily available, feasible feedstock. Little is known about the advantages of using manure as a biomass feedstock. According to a group of researchers in Australia, manure-based biochar "has advantages over typically used plant-derived material because it is a by-product of another industry and in some regions is considered a waste material with little or no value.

It can therefore provide a lower cost base and alleviate sustainability concerns related to using purpose-grown biomass for the process."[16]

Biochar Handling

The spreading of biochar as a soil amendment is ripe for further exploration. Specific questions concern the ideal time to apply biochar and how to ensure that it remains in place once applied and does not cause a risk to human health or degrade air quality.[17] Particulate matter, in the form of dust that is hard for the human body to filter, may be distributed in abnormal quantities if the biochar is mishandled. There are potential public safety concerns for the handling of biochar, as it is a flammable substance. Additionally, the amount of land available for biochar application requires further investigation.

Biochar System Deployment

Biochar systems are designed based on the feedstock to be decomposed and the energy needs of an operation. It would be ambitious to expect a "one size fits all" standard biochar system. According to proponents, a series of mass-produced biochar systems designed for the needs of a segment of the agriculture or forestry communities might prove to be feasible (e.g., forestry community in the southeastern region, corn grower community in the midwestern region, poultry producer community in the mid-Atlantic region). Extensive deployment of biochar systems would depend on system costs, operation time, collaboration with utility providers for the sale of bio-oil, and availability of information about technology reliability.

POLICY CONTEXT

Climate Change Debate

Carbon offsets were a prominent factor in the climate change debate in the 111th Congress. A carbon offset is defined as "a measurable avoidance, reduction, or sequestration of carbon dioxide (CO_2) or other greenhouse gas (GHG) emissions."[18] Carbon sequestration projects are one type of carbon offset. In addition to direct carbon capture and sequestration activities, Congress may consider the role of biological (indirect) sequestration—such as biochar production technology— that can be implemented by agricultural producers at the field level.[19] The establishment of an offset program, identification of eligible project types, carbon offset ownership, and offset verification may be pertinent to the adoption of biochar production technology.[20]

Congress may examine the use of biochar as an indirect carbon sequestration technology that could be used to offset carbon emissions from major emitters. In 2008, 6% of total U.S. greenhouse gas emissions were attributed to the agricultural sector.[21] While not as large as the amounts produced by some other sectors, agricultural emissions come from a large number of decentralized sources, leading many to conclude that controlling such emissions would be difficult. On the other hand, some argue that soil carbon sequestered as a result of biochar application is easily quantifiable and transparent, which may be ideal for carbon trading requirements. Others contend that ancillary benefits could include additional revenue earned by agricultural producers through the sale of carbon credits earned from biochar application or the sale of biochar as a soil amendment. Energy costs for a producer's operation may be reduced by using the energy generated from the biochar production system. Additionally, some assert that the use of biochar results in higher crop yields. This could be a criterion to consider within the larger land use debate.

Legislation Introduced During the 111th Congress

The Water Efficiency via Carbon Harvesting and Restoration (WECHAR) Act of 2009,[22] introduced during the 111th Congress, sought, among other things, to establish loan guarantee programs that would develop biochar technology to use excess plant biomass and establish biochar demonstration projects on public lands. The legislation was primarily focused on woody biomass as the feedstock. Some asserted that the legislation addressed research and development needs for biochar production. Others argued that the legislation lacked specific actions regarding technology transfer or commercial development of biochar production systems.

The discussion draft of the American Power Act offered during the 111th Congress contained three provisions relevant to biochar.[23] The first provision identified projects for biochar production and use as an eligible project type under the domestic offsets program.[24] The second provision instructed the U.S. Environmental Protection Agency (EPA) to submit to Congress a report on the sources and effects of black carbon emissions, and strategies to reduce black carbon emissions, including "research and development activities needed to better characterize the feasibility of biochar techniques to decrease emissions, increase carbon soil sequestration, and improve agricultural production, and if appropriate, encourage broader application of those techniques."[25] The third provision directed the Secretary of Agriculture to provide grants for up to 60 facilities to "conduct research, develop, demonstrate and deploy biochar production technology for the purpose of sequestering carbon."[26]

Farm Bill

The 110th Congress promoted biochar development through the 2008 farm bill (P.L. 110-246), which listed it under grants for High Priority Research and Extension Areas. Noted research areas include biochar production and use, co-production with bioenergy, soil enhancements, and soil carbon sequestration. Listing biochar development as a high-priority research area in the 2008 farm bill did not authorize a specific appropriations amount. Funding for biochar development research would be determined in future appropriation bills and by the U.S. Department of Agriculture. Farm managers facing needs with respect to soil fertility, residue and manure management, energy efficiency, and additional revenue generation may benefit from a policy that further supports biochar production and use (e.g., technology practice standard, cost-share).

Long-Term Prospects

Biochar's fate as a viable component of the long-term solution to mitigate climate change by way of carbon sequestration depends upon further development by the scientific and technology transfer communities. In particular, biochar's practical application at various locations and scales using multiple feedstocks throughout the United States is an area for additional study. Policy that encourages academia and other institutions to conduct in-depth research and development could quicken the pace of technology deployment. One study hypothesizes that a U.S. research program (technical demonstrations and a coordinated national field trial program), funded at $100 million-$150 million for eight years, could support the research, development, and deployment of commercial-scale biochar production and use.[27]

An assessment of external factors (e.g., feedstock transportation costs and disposal fees) associated with the economic growth of biochar production systems, similar to studies conducted for the biofuels industry, could provide guidance on the types of federal financial and technical incentives necessary to spur development (e.g., regulatory requirements, technical standards). The definition of biomass used in climate change and energy legislation will directly affect the eventual impact of biochar in limiting GHG emissions.[28] Indeed, the biomass definition would determine what sources of material are deemed acceptable and which lands would be eligible lands for biomass removal.

U.S. Department of Agriculture Activities

According to a U.S. Department of Agriculture (USDA) Agricultural Research Service (ARS) official, an estimated $2.6 million was spent in 2010 on in-house biochar research by ARS.[29] ARS has multiple projects underway to ascertain biochar's value as a soil amendment given varying soil, climate, and cropping system conditions. Additionally, ARS is examining feedstock types and pyrolysis conditions for biochar production. ARS estimates that the United States could use biochar to sequester 139 Tg of carbon on an annual basis if it were to harvest and pyrolyze 1.3 billion tons of biomass.[30]

End Notes

[1] A series of presentations delivered at the 2008 United Nations Climate Change Conference elevated interest in biochar as an immediate response to mitigate climate change, given its carbon sequestration ability. “Dangerous Sea Level Rise Imminent without Large Reductions of Black Carbon and Implementation of Other Fast-Action Mitigation Strategies,” Environmental Research Web, December 12, 2008.

[2] Institute for Governance & Sustainable Development and IGDS/INECE Climate Briefing Note, “Significant Climate Mitigation Is Available from Biochar,” December 8, 2008.

[3] Syngas, or synthesis gas, consists of varying concentrations of carbon monoxide and hydrogen and can be used as a replacement to natural gas or liquid fuel. Bio-oil is a liquid fuel that can be used, once upgraded, as a motor fuel.

[4] A soil amendment improves the physical properties of soil (e.g., moisture-holding capacity, nutrient retention ability).

[5] Johannes Lehmann, John Gaunt, and Marco Rondon, “Bio-char Sequestration in Terrestrial Ecosystems—A Review,” Mitigation and Adaptation Strategies for Global Change, vol. 11 (2006), pp. 403-427.

[6] Bruno Glaser, Johannes Lehmann, and Wolfgang Zech, “Ameliorating Physical and Chemical Properties of Highly Weathered Soils in the Tropics with Charcoal—A Review,” Biology and Fertility of Soils, vol. 35 (2002), pp. 219-230.

[7] Emma Marris, “Black Is the New Green,” Nature, vol. 442, no. 10 (August 2006), pp. 624 626.

[8] For more information on nitrous oxide emissions, see CRS Report R40874, Nitrous Oxide from Agricultural Sources: Potential Role in Greenhouse Gas Emission Reduction and Ozone Recovery, by Kelsi Bracmort.

[9] For more information on methane emissions from agricultural sources, see CRS Report R40813, Methane Capture: Options for Greenhouse Gas Emission Reduction, by Kelsi Bracmort et al.

[10] Johannes Lehmann, “Bio-energy in the Black,” Frontiers in Ecology and the Environment, vol. 5, no. 7 (2007), pp. 381-387.

[11] Tyler Hamilton, “The Case for Burying Charcoal,” Technology Review, April 26, 2007.

[12] Yosuke Yanai, Koki Toyota, and Masanori Okazaki, “Effects of Charcoal Addition on N2O Emissions from Soil Resulting from Rewetting Air-Dried Soil in Short-Term Laboratory Experiments,” Japanese Society of Soil Science and Plant Nutrition, vol. 53 (2007), pp. 181-188.

[13] Johannes Lehmann, “A Handful of Carbon,” Nature, vol. 447 (May 10, 2007), pp. 143-144.

[14] Johannes Lehmann, “Bio-energy in the Black,” Frontiers in Ecology and the Environment, vol. 5, no. 7 (2007), pp. 381-387.

[15] Stephan M. Haefele, “Black Soil, Green Rice,” Rice Today, April-June 2007, p. 27.

[16] K. Y. Chan, L. Van Zwieten, and I. Meszaros et al., “Using Poultry Litter Biochars as Soil Amendments,” Australian Journal of Soil Research, vol. 46 (2008), pp. 437-444.

[17] David A. Laird, “The Charcoal Vision: A Win-Win-Win Scenario for Simultaneously Producing Bioenergy, Permanently Sequestering Carbon, While Improving Soil and Water Quality,” Agronomy Journal, vol. 100, no. 1 (2008), pp. 178-181.

[18] For more information on carbon offsets, see CRS Report RL34241, Voluntary Carbon Offsets: Overview and Assessment, by Jonathan L. Ramseur.

[19] For more information on carbon capture and sequestration, see CRS Report RL33801, Carbon Capture and Sequestration (CCS), by Peter Folger; and CRS Report RL33898, Climate Change: The Role of the U.S. Agriculture Sector, by Renée Johnson.

[20] For more information on the role of offsets in current greenhouse gas legislation, see CRS Report R40643, Greenhouse Gas Legislation: Summary and Analysis of H.R. 2454 as Passed by the House of Representatives, coordinated by Mark Holt and Gene Whitney.

[21] U.S. Environmental Protection Agency, 2010 U.S. Greenhouse Gas Inventory Report, EPA 430-R-10-006, April 2010, http://www.epa.gov/climatechange/emissions/usinventory report.html.

[22] S. 1713.

[23] Discussion draft available at http://kerry.senate.gov/americanpoweract/pdf/APAbill.pdf.

[24] American Power Act (discussion draft), Title VII, Part D, Sec. 734.

[25] American Power Act (discussion draft), Title VII, Subtitle C, Part II, Sec. 2211.

[26] American Power Act (discussion draft), Title VII, Subtitle C, Part II, Sec. 2214.

[27] Stephen Brick and Sasha Lyutse, Biochar: Assessing the Promise and Risks to Guide U.S. Policy, Natural Resources Defense Council, November 2010, http://www.nrdc.org/energy/files/biochar_paper.pdf.

[28] For more information on biomass definitions, see CRS Report R40529, Biomass: Comparison of Definitions in Legislation Through the 111th Congress, by Kelsi Bracmort and Ross W. Gorte.

[29] Correspondence with USDA ARS official, January 2011.

[30] Correspondence with USDA ARS official, January 2009. A teragram (Tg) is equivalent to 1 trillion grams. A Tg C (teragram of carbon) is a unit of measurement used to compare greenhouse gases emitted from different sources on the same basis.

In: Carbon Considerations
Editors: W. Green and L. Wayman ISBN 978-1-62618-435-0

Chapter 4

BIOENERGY PRODUCTION SYSTEMS AND BIOCHAR APPLICATION IN FORESTS: POTENTIAL FOR RENEWABLE ENERGY, SOIL ENHANCEMENT, AND CARBON SEQUESTRATION*

Kristin McElligott, Deborah Page-Dumroese and Mark Coleman

ABSTRACT

Bioenergy production from forest biomass offers a unique solution to reduce wildfire hazard fuel while producing a useful source of renewable energy. However, biomass removals raise concerns about reducing soil carbon and altering forest site productivity. Biochar additions have been suggested as a way to mitigate soil carbon loss and cycle nutrients back into forestry sites; yet, little is known about the effects of intentional biochar amendments to temperate forest soil in conjunction with biomass removals for bioenergy production. In this review, we evaluate the potential for mobile bioenergy systems and the environmental implications of biochar application in forests. Using forest biomass that

* This research note (RMRS-RN-46) was released by the USDA, Forest Service, Rocky Mountain Research Station, October 2011.

accumulates annually during forest harvest operations, bioenergy can be produced on-site and the biochar that is generated can be redistributed to return nutrients and help improve water holding capacity of the site. Little is known about the short- and long-term impacts of biochar application in forest ecosystems. Some sites may benefit from biochar application, while others show no or negative responses. Field studies on soil and vegetation responses combined with laboratory studies will elucidate the best sites for biochar application and sustainable bioenergy production.

THE NEED FOR BIOMASS REMOVALS

Removing residual woody biomass from forest management activities is important for hazardous fuel reduction and forest health improvement. Improper fire management during the past century has resulted in overstocked forests or excess coarse woody debris on the soil surface (Kauffman 1990). An estimated 73 million acres of National Forest land in the western United States have been identified as having unnatural or excessive amounts of woody biomass (USDA Forest Service 2003). Residues associated with timber harvesting are also significant, which further results in increased susceptibility to catastrophic fire. As a consequence of increasing wildfire occurrence and overstocked stands that alter forest health, land managers have begun to thin trees and remove residues. Removal operations are often mandated by law for both public and private land owners, but financial support and incentive is limited with these operations (Healthy Forests Initiative 2003).

In the western United States, the cost of biomass removal often exceeds its value, despite increasing interest in forest biomass utilization. Burgeoning interest in using woody biomass for heat or bioenergy is a result of rising fuel costs, greenhouse gas emissions from fossil fuels, and the threat of stand-replacing wildfires; however, the collection and transportation of woody debris and harvesting waste from forests are among many economic impediments to woody biomass utilization.

There is little to no market for residual biomass, deeming it the lowest value material removed from the forest (Evans 2008). Consequently, it is rarely a source of income for land owners. Costs of low-grade wood such as forestry residues range from \$0.10 to \$40 per ton for chips. The median cost for removal projects that did not result in profits was \$625 per acre, but it reached up to \$1000 per acre in the western United States for mechanical clearing alone (Rummer and others 2003). In 2005, the median cost of bringing biomass to the roadside was \$680 per acre for mild slopes (USDA

Forest Service 2005) not including costs for haul distances. The U.S. Department of Agriculture, Forest Service (2005) determined the western forest break-even hauling distance to be a maximum of 86 miles, assuming a price of $30 per dry ton delivered to the mill for chips and chip transport costs of $0.35 per dry-ton-mile, excluding treatment costs. Biomass removal costs are highly variable depending on stand conditions, locations, and markets (Lynch and Mackes 2003), making it difficult to estimate standard costs for these operations. Minimizing costs of collection and delivery of biomass to end-users is essential to effectively and economically use this resource.

An important, emerging market for woody biomass is energy production. Woody biomass is a local, renewable resource that can be used for transportation fuel, heat, and power. Additionally, if used as a sustainable substitute for fossil fuels, it can reduce greenhouse gas emissions. Utilizing residues from logging, pre-commercial thinning, and hazardous fuel reductions for energy production will help meet U.S. energy independence goals while promoting forest stand health and reducing wildfire risks. It is estimated that in the 15 western United States, more than 28 million acres of forestlands could benefit from hazardous fuel reduction treatments, yielding approximately 345 million oven-dry tons from accessible areas (Rummer and others 2003). Yet, due to the low value of biomass, limited accessibility, and varying biomass markets, the majority of residues are left to decay at the site or are incinerated in slash piles, which is a waste of potential energy. Piling and burning slash redistributes nutrients on the site, or concentrates them to localized areas, which may lead to lower average site productivity (Binkley 1986). Further, slash burning releases pollutants, including greenhouse gasses, into the atmosphere and can result in lost site nutrients. Volatile elements such as carbon (C), nitrogen (N), and sulfur (S) are readily released as gases, and some C and N is lost from the ecosystem (Hosking 1938; Knight 1966; Tiedemann 1987; Caldwell and others 2002). Phosphorus (P) can also be lost, but in lower quantities than N and S. These nutrients are frequently limiting in forest environments (Kishchuk and Brockley 2002; Fox and others 2007); therefore, it is important to retain onsite nutrient stores instead of causing losses from volatilization. Consequently, slash burning is an unwise method for biomass removal.

Biomass removal projects raise concern over ecological impacts. Biomass typically consists of mostly fine-diameter material with high nutrient content. The concerns of consequential ecological impacts with removals are due, in part, to the high nutrient content in tree tops and limbs and to the lack of research to evaluate site-specific, short- and long-term ecological effects of removals. While the site impacts are thought to be low, there is considerable concern that an established bioenergy market would degrade site nutrient stocks over time, and over exploitation of this resource could be a negative consequence of biomass harvesting for energy production (Kimmins 1997). The dead wood left behind from harvest residues would otherwise decay and slowly recycle nutrients back to the soil and forest (Johnson and Curtis 2001; Mahendrappa and others 2006). It is understood that bole-only removal during timber harvesting has little impact on the growth of succeeding forest stands; however, whole-tree removals on nutrient-poor sites have had negative impacts (Kimmins 2004). Further, it is understood that disturbing or removing the surface organic horizons and displacing soil may have significant negative impacts on the subsequent stand rotations (Fleming and others 2006). There exists little to no field research on the impacts of removing small-diameter biomass material (e.g., twigs, small branches, and needles that contain high concentrations of nutrients), making long-term impacts associated with these removals difficult to infer. Potential removal consequences will likely depend on the initial site quality and soil properties, the frequency and intensity of harvests, and the ability of the site to replace nutrients between removals (Kimmins 2004). Although forest ecosystems are resilient and can maintain large stocks of nutrients, increasing the frequency of biomass removal may exceed the natural capacity for nutrient replenishment between removals, emphasizing the need to understand the implications of biomass removal.

MOBILE FAST-PYROLYSIS

A sustained bioenergy production system might include removing the energy but not the nutrients or returning the nutrients after energy is extracted from the biomass. Innovative technology allows for in-woods conversion of biomass to higher value energy products through pyrolysis, with an opportunity to retain the nutrient status at removal sites. Mobile fast- pyrolysis bioenergy production systems (Badger and Fransham 2006) may be one approach to profitable and sustainable biomass utilization. These units can be located at or near biomass removal locations and are capable of converting bulky, low-value biomass into an easily stored and transportable fuel (bio-oil) that can be effectively used for heat, power, and chemical production (Garcia-Perez and others 2007). Additionally, these units produce a charcoal byproduct (biochar) that has market value of its own, but might best be used by returning it to the site of energy extraction as a soil amendment and as a means of soil C sequestration. Such an approach has been implemented in agricultural systems (Lehmann and others 2006; Laird 2008) but is particularly attractive in forest ecosystems where the biochar can be both produced and immediately returned to the site of energy extraction.

A mobile pyrolysis unit could provide an economical hazardous fuels reduction system by producing alternative sources of energy that can be sold to offset biomass removal costs. Mobile pyrolysis units are designed to convert biomass into bio-oil, biochar, and syngas through thermal decomposition of organic material under anaerobic conditions (Boucher and others 2000; Bridgewater 2004). Pyrolysis involves rapidly heating the biomass (500 °C/sec) to moderate reaction temperatures (400 to 600 °C) followed by rapid cooling with short vapor residence times (1 to 2 s). Biomass feedstock is dried to less than 10% water in order to minimize the water in the resulting liquid product (Bridgewater 2004; Mohan and others 2006). Pyrolysis produces 60 to 75% w/w bio-oil, 15 to 25% w/w solid biochar, and 10 to 20% w/w non-condensable gases, but exact proportions are largely dependent on the feed-stock and process temperatures (Mohan and others 2006). No waste, other than flue gas and ash, is generated in the conversion process as the bio-oil and biochar can each be used as a fuel and the synthesis gas can be recycled back into the pyrolysis process, yielding an energy output and making the process sustainable (Bridgewater 2004; Mohan and others 2006). Such an energy production system would help maintain or enhance site productivity and mitigate nutrient depletion through the application of biochar.

BIOCHAR SOIL APPLICATION

Biochar, a byproduct of the pyrolysis process, is biomass-derived black C. It is analogous to charcoal manufactured through traditional or modern pyrolysis methods and to black C found naturally in fire-prone ecosystems. Biochar has been used as a soil amendment to improve soil nutrient status, C storage, and/or filtration of percolating soil water (Lehmann and Joseph 2009), primarily in agricultural systems. Biochar from pyrolysis and charcoal produced through natural burning share key characteristics, including long residence time in soils and a soil conditioning effect (Glaser and others 2002). Biochar has an inherent energy value that can be used to maximize the energy output of pyrolysis. However, research has shown that applying biochar to soil may be more desirable as it can increase soil organic carbon (SOC) and improve the supply of nutrients to plants, therefore enhancing plant growth and soil physical, chemical, and biological properties (Glaser and others 2002; Lehmann and others 2002; Rondon and others 2007). Regardless of its commercial market value, biochar presents an opportunity to return site nutrients lost from biomass removal projects, which may overshadow other potential uses.

Biochar Composition

Biochar is produced from biomass and is predominantly composed of recalcitrant organic C with contents of plant micro- and macro-nutrients retained from the original feedstock. We know from research on wildfire occurrence and the development of Anthrosols (e.g., terra preta soils) in the Amazon that charcoal can remain in the soil for hundreds to thousands of years (Agee 1996; Lehmann and Rondon 2006). Consequently, biochar can rapidly increase the recalcitrant soil C fraction of soil. The C in biochar is held in aromatic form, which is resistant to decomposition when added as a soil amendment (Amonette and Joseph 2009), making it a C sequestration tool. However, composition varies by feedstock type and conditions of pyrolysis (Downie and others 2009). Actual C contents can range between 172g kg^{-1} and 905g kg^{-1}. Nitrogen content ranges from 1.8 kg^{-1} to 56.4g kg^{-1}, total P from 2.7g kg^{-1} to 480g kg^{-1} and total potassium (K) from 1.0g kg^{-1} to 58g kg^{-1} (Lehmann and others 2003; Lima and Marshall 2005; Chan and others 2007). Biochar also contains varying concentrations of other elements such as oxygen (O), hydrogen (H), S, base cations, and heavy metals (Goldberg 1985; Preston and Schmidt 2006). Freshly produced biochar consists of a crystalline phase with graphene layers and an amorphous phase of aromatic structures (Lehmann and others 2005; Cohen-Ofri and others 2007). The outer surfaces contain various O and H functional groups, and the graphene sheets may contain O groups and free radicals (Bourke and others 2007). Additionally, biochar has been produced with a range of pH values between 4 and 12, but the final pH is dependent upon the original feedstock and operating conditions (Lehmann 2007). Generally, low pyrolysis temperatures (<400 °C) yield acidic biochar, while increasing pyrolysis temperatures produce alkaline biochar. Once incorporated to the soil, surface oxidation occurs due to reactions of water, O_2, and various soil agents (Cheng and others 2006; Lehmann 2007). The cation exchange capacity (CEC) of fresh biochar is typically very low but increases with time as the biochar ages in the presence of O_2 and water (Cheng and others 2006, 2008; Liang and others 2006).

There are increasing concerns that contaminants are being retained in biochar and may leach into the soil once added as an amendment; however, contaminant type and quantity are dependent on the origin of the pyrolysis feedstock and the conversion process. Biochar may contain contaminants such as heavy metals and organic compounds, but these are commonly associated with sewage sludge and treated wood feedstocks (Lievens and others 2009), and would likely not be an issue if biochar is produced from forest biomass. Contaminants contained in feedstocks could undergo changes during the pyrolysis process and be destroyed or transformed into benign compounds, while others could be retained in the biochar and be potentially detrimental if added to the soil. In addition, some contaminants (e.g., polycyclic aromatic hydrocarbons) can be formed during pyrolysis. Polycyclic aromatic hydrocarbons (PAH) can be formed from any carbonaceous feedstock, but concentrations are feedstock dependent (Zhurinsh and others 2005). Thus, prior to large-scale application, it is important to understand the chemical composition of the initial feedstock and biochar to avoid potential negative environmental consequences.

Impacts of Biochar on Soil

Biochar offers substantial potential for soil improvement because of its unique physical, chemical, and biological properties and their interactions with soil and plant communities. If used as a soil amendment, biochar could mitigate the possible negative impacts of forest biomass removal operations. However, uncertainties surround the potential short- and longterm effects of intentional biochar application in many regions and ecosystems (namely temperate forests) as most evidence comes from agricultural systems. While additions have largely been neutral or positive (reviewed by Sohi and others 2010), negative impacts are possible, demonstrating the need for a comprehensive understanding of biochar's origin, production, and functional properties.

The physical properties of biochar offer several soil benefits. Its highly porous nature is a result of the cell wall structure of the biomass feedstock. A wide range of pore sizes within the biochar results in low bulk density and high surface area material. Biochar incorporation can alter soil physical properties such as structure, pore size distribution, and density, and has implications for soil aeration, water holding capacity, plant growth, and soil workability (Downie and others 2009). Evidence suggests that biochar application into soil may increase the overall net soil surface area (Chan and others 2007) and, consequently, may improve soil water and nutrient retention (Downie and others 2009) and soil aeration, particularly in fine-textured soils (Kolb and others 2007). Biochar has a bulk density much lower than that of mineral soils (~0.3 Mg m^{-3} for biochar compared to typical soil bulk density of 1.3 Mg m^{-3}); therefore, application can reduce the overall total bulk density of the soil, which is generally desirable for most plant growth (Brady and Weil 2004).

Increased surface area and porosity and lower bulk density in mineral soil treated with biochar can alter water retention and aggregation and can decrease soil erosion (Piccolo and Mbagwu 1990; Piccolo and others 1996; Mbagwu and Piccolo 1997). Soil water retention is determined by the distribution and connectivity of pores in the soil matrix, which are largely affected by soil texture, aggregation, and organic matter content (Brady and Weil 2004). Biochar has a higher surface area and greater porosity relative to other types of soil organic matter and can therefore improve soil texture and aggregation, which improves water retention. Biochar's initial physical properties occur at a range of scales and affect the proportion of water than can be retained. Kishimoto and Sugiura (1985) estimated the inner surface area of biochar formed between 400 and 1000 °C to range from 200 to 400 m^2g^{-1}. Van Zwieten and others (2009) reported the surface area of biochar that was derived from papermill waste with slow pyrolysis to be 115 m^2g^{-1}. These properties are expected to change over time with physical weathering but have not been explicitly examined, resulting in uncertainties associated with the longevity of these beneficial physical changes in soil.

Soil moisture retention is an indirect result of altered soil aggregation and structure after biochar application (Brodowski and others 2006). Biochar can affect soil aggregation through interactions with soil organic matter, minerals, and microorganisms; however, the surface charge characteristics and their development over time determine the long-term effect on soil aggregation. Glaser and others (2002) reported that Anthrosols that were enriched with charcoal had surface areas three times higher than those of surrounding Oxisols and had an increased field capacity of 18%. Tryon (1948) studied the effect of charcoal on the percentage of available moisture in soils of different textures and found different response among soils. In sandy soil, available moisture increased by 18% after adding 45% biochar by volume, while no changes were observed in loamy soil, and soil available moisture decreased in the clayey soil. The high surface area of biochar can lead to increased water retention, although the effect seems to depend on the initial texture of the soil. Improved water holding capacity with biochar addition is most commonly observed in coarse-textured or sandy soils (Glaser and others 2002; Gaskin and others 2007). The impact of biochar additions on moisture content may be due to increased surface area relative to that found in coarse-textured soils (Glaser and others 2002). Therefore, improvements in soil water retention by biochar additions may only be expected in coarse-textured soils or soils with large amounts of macropores. Additionally, a large amount of biochar may need to be applied to the soil before it increases water retention.

Biochar has the potential to increase nutrient availability for plants (Lehmann and others 2002). Nutrient availability can be affected by increasing cation exchange capacity, altering soil pH, or directly adding nutrients from biochar. One potential mechanism for enhanced nutrient retention and supply following biochar amendment is increasing CEC by up to 50% as compared to unamended soils (Tryon 1948; Mbagwu and Piccolo 1997; Lehmann and others 2002; Liang 2006). Biochar has a greater ability to adsorb and retain cations in an exchangeable form than other forms of soil organic matter due to its greater surface area and negative surface charge (Liang and others 2006). Studies have shown significant increases in the availability of all major cations (Glaser and others 2002; Lehmann and others 2002; Topoliantz and others 2005). Tryon (1948) found increasing amounts of exchangeable bases in sandy and loamy soils after adding 45% hardwood and conifer charcoals. Additionally, freshly produced biochar is reported to have an anion exchange capacity. Cheng and others (2008) found biochar to exhibit an anion exchange capacity at pH 3.5, which decreased to zero over time as the biochar aged in soil.

Biochar has a higher sorption affinity for a range of organic and inorganic compounds and a higher nutrient retention ability compared to other forms of soil organic matter (Bucheli and Gustafsson 2000, 2003; Allen-King and others 2002; Kleineidam and others 2002; Nguyen and others 2008). Once biochar is added to the soil, abiotic and biotic surface oxidation of biochar results in increased surface carboxyl groups, a greater negative charge, and subsequently, an increasing ability to sorb cations (Cheng and others 2006, 2008). Treated soil also exhibits an ability to sorb polar compounds, including many environmental contaminants (Yu and others 2006). Cation exchange capacity of biochar is highly variable depending upon the pyrolysis conditions under which it is produced. Cation exchange capacity is lower at low pyrolysis temperatures and significantly increases when produced at higher temperatures

(Lehmann 2007). Freshly produced biochars have little ability to retain cations, resulting in minimal CEC (Cheng and others 2006, 2008; Lehmann 2007), but this increases with time in soil with surface oxidation (Cheng and others 2006). These results support the findings of high CEC observed in Amazonian Anthrosols (Liang and others 2006).

Biochar can serve as a liming agent, resulting in increased pH and nutrient availability for a number of different soil types (Glaser and others 2002; Lehmann and Rondon 2006). The carbonate concentration of biochar facilitates liming in soils and can raise pH of neutral or acidic soil (Van Zweiten and others 2007). Mbagwu and Piccolo (1997) reported increases in pH of various soils and textures by up to 1.2 pH units from pH 5.4 to 6.6. Tryon (1948) reported a greater increase in pH in sandy and loamy soils than in clayey soils. The pH of various soils increased more after applications of hardwood charcoals (pH 6.15) than of conifer charcoals (pH 5.15), probably because of different ash contents of 6.38% and 1.48%, respectively (Glaser and others 2002).

Biochar feedstock and pyrolysis conditions largely determine the resulting carbonate concentrations, making some biochar a better liming agent than others. Concentrations of carbonates can vary from 0.5 to 33% (Chan and others 2007) depending on starting conditions. Hardwood charcoals are reported to have substantial carbonate concentrations and prove more effective in reducing soil acidity and, therefore, have a larger influence on soil fertility (Steiner and others 2007). The liming of acidic soils decreases aluminum saturation, while increasing cation exchange capacity and base saturation (Cochrane and Sanchez 1980; Mbagwu and Piccolo 1997; Fisher and Binkely 2000). Additionally, nutrient availability may actually increase beyond the amount anticipated by cation exchange sites as a result of the soluble salts available in the biochar.

The liming effect associated with biochar may not be ideal for all soil types and plant communities. Increased soil pHs associated with biochar additions have caused micronutrient deficiencies in agricultural crops (Kishimoto and Sugiura 1985) and forest vegetation (Mikan and Abrams 1995); thus, it is important to acknowledge the presence of calcifuge vegetation prior to application. In addition, many forest plants, fungi, and bacteria thrive in lower pH soils (Meurisse 1976, 1985); therefore, altering forest soil pH through the addition of biochar may result in unfavorable shifts in above- and belowground flora. Understanding interactions among biochar production and application conditions and soil texture, organic matter, and pH will be key in determining long-term effects of biochar application on forest soils.

In the short term, biochar may supply a source of plant-available nutrients once applied to the soil (Gaskin and others 2008; Sohi and others 2010). A small fraction of nutrients in the feedstock, apart from N, are retained in biochar in a potentially extractable form. It is uncertain whether these soluble nutrients are released instantaneously once added to the soil environment or if they are released over time (Sohi and others 2010); this will likely depend on the starting soil physical properties. The rapid introduction of readily available nutrients and small amounts of labile C retained in biochar could promote mineralization of soil organic matter (Wardle and others 2008a), especially in nutrient-limited environments. Additionally, alkaline biochar may increase the pH of acidic soils, subsequently stimulating microbial activity thereby further promoting mineralization or decomposition of existing soil organic matter.

Biochar properties may enhance soil microbial communities and create microenvironments that encourage microbial colonization. Biochar's pores, high internal surface area, and increased ability to adsorb organic matter provide a suitable habitat to support soil microbiota, which catalyze processes that reduce N loss and increase nutrient availability for plants (Winsley 2007). It has been suggested that pores serve as a refuge to microbes by protecting them from predation and desiccation, while the organic matter adsorbed to biochar provides C energy and mineral nutrient requirements (Saito and Muramoto 2002; Warnock and others 2007). In temperate ecosystems with wildfire-produced charcoal, N mineralization and nitrification are enhanced (Berglund and others 2004; Gundale and DeLuca 2007) by creating favorable microenvironments that enhance microbial colonization (Pietikäinen and others 2000; Warnock and others 2007). If microbial activity is able to oxidize biochar, we need to know which microbes can achieve this, the mechanism by which it occurs, and under what conditions and rates it takes place.

Evidence supporting enhanced microbial abundance and the buildup of recalcitrant soil C comes from studying charcoal-amended Anthrosols and wildfire charcoal. While many studies suggest biochar additions are beneficial for increasing microbial activity and C storage, others have reported accelerated decomposition of soil organic matter (priming) after fresh biochar or charcoal additions. Liang and others (2010) reported high stabilization of organic material added to soils from a tropical environment containing aged charcoal. They reported 25.5% less mineralization of added organic matter to Anthrosols compared to unamended adjacent Oxisols. While the charcoal-amended Anthrosol had more than two times the amount of microbial biomass as adjacent soils, CO_2 respiration was lower compared to unamended adjacent soils. This suggests that the microbial biomass associated with charcoal additions has higher metabolic efficiency (Liang and others 2010). Similar findings supporting microbial proliferation and decreased soil respiration have been reported in mineral soil that was amended with varying rates of maize-derived biochar (Jin and others 2008). Conversely, the potential for biochar to cause or accelerate decomposition of soil surface organic horizons (humus) was reported in a 10-year study of litter bags in the boreal zone (Wardle and others 2008a), where a more rapid loss of humus in the presence of charcoal was demonstrated.

Similarly, Steinbeiss and others (2009) showed that homogeneous biochars with or without N could stimulate the loss of soil organic C (between 8 and 13%) in both agricultural and forest soils. Evidence also suggests that the availability of soil N is a controlling factor for the priming effect of char (Neff and others 2002; DeLuca and others 2006; Gundale and DeLuca 2006). Whether biochar application stabilizes soil organic matter and soil C or it results in priming is still speculated and warrants further investigation (Lehmann and Sohi 2008; Wardle and others 2008a, 2008b; Sohi and others 2010).

Plant Growth Response to Biochar Additions

Biochar can be used as a soil amendment to improve soil quality and crop productivity in a variety of soils (Blackwell and others 2009). This has been demonstrated primarily in soils that are highly weathered or degraded through agricultural activities (Glaser and others 2002; Kimetu and others 2008). Much of the initial information concerning biochar effects on soil parameters and crop yields has come from studying properties of Amazon Dark Earth Anthrosols (terra preta; Laird and others 2009). Soils in this region were created by pre- Columbian Indians (Smith 1980; Woods and McCann 1999) using a slash-and-char method. Compared to the surrounding Oxisols, these Anthrosols are characterized as having enhanced levels of soil organic matter, CEC, pH, base saturation, and nutrients such as N, P, K, and calcium (Sombroek 1966; Smith 1980; Sombroek and others 1993; Glaser and others 2001; Lehmann and others 2003; Liang and others 2006). Additionally, the improved nutrient retention and enhanced soil fertility of these Anthrosols has resulted in higher crop yields relative to the adjacent Oxisols (Lehmann and others 2002; Liang and others 2006; Solomon and others 2007). Their nutrient content, dark color, and greater fertility are partially attributed to their high biochar (charcoal) content (Glaser and others 2001). These soils have C contents of up to 150 g C kg^{-1} as compared to surrounding Oxisols that have 20 to 30 g C kg^{-1} (Sombroek 1966; Lehmann and others 2002).

As a result of the greater fertility of these Anthrosols, numerous greenhouse and field trials have been implemented to evaluate impacts of fresh biochar on crop biomass yield and soil properties; however, the majority of the reported studies have taken place in tropical environments, resulting in little understanding of biochar potential in temperate regions. In a pot experiment, Lehmann and others (2002) found that biochar increased rice biomass by 17% and cowpea biomass by 43% when applied at rates of 68t C ha^{-1} to 135t C ha^{-1}. This growth was attributed to direct nutrient additions from biochar of P, K, and Copper. Other studies have attributed positive plant growth to positive changes in soil biogeochemistry resulting from biochar additions (Iswaran and others 1980; Wardle and others 1998; Hoshi 2001; Lehmann and others 2002; Chan and others 2007; Van Zwieten and others 2007). For example, Iswaran and others (1980) reported a 51% increase in biomass in soybean crops with biochar additions of 0.5t ha^{-1} in a pot study. In a field experiment in Japan, Hoshi (2001) found a 20% increase in volume and 40% increase in height of tea trees after biochar additions. In both field and pot studies in Zambia, Chidumayo (1994) reported better seed germination (30% enhancement), shoot height (24%), and biomass production (13%) among seven native woody plants on soils under charcoal kilns as compared to undisturbed Alfisols and Ultisols. Additionally, in tropical zones, larger yield increases have been reported when biochar was applied in combination with inorganic or organic fertilizer (Glaser and others 2002; Lehmann and others 2002; Chan and others 2007; Van Zwieten and others 2007; Steiner and others 2007), with increases of 200% reported relative to unamended, unfertilized plots (Yamato and others 2006). A combination of the abilities to raise soil pH (Hoshi 2001; Yamato and others 2006; Rondon and others 2007; Van Zwieten and others 2007), improve physical properties (e.g., water holding capacity; Iswaran and others 1980), and retain soil nutrients and reduce leaching losses (Hoshi 2001; Lehmann and others 2003; Lehmann 2007) are likely reasons that biochar has contributed to increased plant productivity.

However, not all biochar effects on soil properties are positive, and declines in plant growth have also been reported with biochar additions. Kishimoto and Sugiura (1985) reported biochar additions at 5 t ha^{-1} decreased soybean yields by 37%, while 15 t ha^{-1} decreased yields by 71%. In southeastern Pennsylvania, Mikan and Abrams (1995) found negative response of vegetation in older (more than 100-year-old) charcoal hearth areas due to presence of charcoal. Tree density and basal area were reduced by 40% in charcoal hearth locations compared to non-hearth areas. Although ancient Amazonian Anthrosols have more favorable characteristics than heavily weathered Oxisols from which they were derived, fresh biochar amendments do not consistently improve soil conditions (Chan and Xu 2009).

Positive plant growth and nutrient content responses to biochar have been commonly observed in association with fertilizer application, while neutral or negative responses have been observed following biochar-only amendments (Yamato and others 2006; Gundale and DeLuca 2007; Asai and others 2009; Blackwell and others 2009; Gaskin and others 2010). Greater plant growth yields can be attributed to changes in soil physical properties previously mentioned (Chan and Xu 2009), but also to the ability of biochar to retain fertilizer nutrients and reduce leaching losses (Lehmann and others 2002). Furthermore, nutrient retention in soils amended with biochar may be attributed to the sorptive capacity of fresh biochar through charge or covalent interactions (Major and others 2009).

However, fertilizer additions are not always capable of ameliorating negative growth responses of fresh biochar additions (Asai and others 2009). It has been suggested that biochar causes N immobilization and could potentially cause N deficiency in plants when applied to soil alone due to high C:N ratios (Sullivan and Miller 2001; Chan and Xu 2009; Lehmann and Joseph 2009), leading to further uncertainty regarding its effect on plant growth. Additions of organic matter with available C:N ratios above 20 are known to cause microbial N immobilization (Fisher and Binkley 2000). Because biochar has a high C:N ratio (up to 400), it is likely that rapid mineralization of a labile C fraction could contribute to reduced soil mineral N, potentially reducing plant available N. However, total C and N content in biochar do not reflect the actual availability of these elements for microbes to cause immobilization. The recalcitrant nature of biochar suggests that few of its components would contribute to immobilization; however, biochar may also sorb organic molecules that have high C:N from soil solution and increase mineralization (Gundale and DeLuca 2007).

Fresh biochar both directly and indirectly influences soil nutrient availability (Blackwell and others 2009; Chan and Xu 2009), which can impact plant growth. Direct effects are largely associated with the retained feedstock nutrients in biochar and are apparent when soil nutrients, plant production, and foliar nutrient concentrations are enhanced with biochar applications (Lehmann and others 2002; Gaskin and others 2010). Amendments of biochar can indirectly affect nutrient availability by adding chemically active surfaces that modify the dynamics of soil nutrients or facilitate soil reaction, modify physical properties of the soil (e.g., reduce bulk density, increase porosity, and increase water holding capacity; Iswaran and others 1980), and encourage the formation of mineral and microbial associations with biochar particles (Pietikäinen and others 2000; Warnock and others 2007). Biochar typically increases pH of acidic soils (Lehmann and others 2002; Gaskin and others 2010; Van Zwieten and others 2010) due to the liming capacity of associated carbonate salts retained in the ash component of biochar. As previously mentioned, this can improve the availability of some nutrients, which is commonly thought to be responsible for positive plant growth responses to biochar amendments (Chan and Xu 2009). However, because it can be difficult to differentiate among direct and indirect effects of biochar application, suffice it to say the combination is largely responsible for nutrient supply responses. Further research is needed to understand short-, mid-, and long-term effects on immobilization and mineralization in conjunction with biochar additions to field environments. We need additional research in order to understand the impacts of application rate, feedstock type, and various site types, especially in forests and temperate regions where data are limited. Amending soils with biochar from various feedstocks will result in differing effects on soil properties and subsequent effects on plant growth. The temperature and heating rate of the pyrolysis process also has important effects on the physical and chemical attributes of the biochar produced (Amonette and Joseph 2009; Downie and others 2009), which will impact soil properties (Gaskin and others 2008). Feedstock such as poultry manure can result in biochar with high pH and P content, while sewage sludge can result in biochar with high N and heavy metal concentrations. Fresh vegetation, wood, or bark may create biochar with neutral pH and nutrient concentrations that reflect feedstock concentrations (Chan and Xu 2009). Gaskin and others (2010) compared biochar derived from peanut shells or wood chips and found peanut-shell biochar had higher nutrient concentrations and raised the pH and base cation concentrations when added to the soil, while wood chip derived biochar had little effect on these parameters. From the limited data available, no optimum range or type of biochar application has been determined to enhance plant productivity (Glaser and others 2002; Lehmann and others 2002). It is likely that the optimum rate of biochar application will depend on each soil type and target plant species.

Biochar Stability and C Sequestration Potential

The long residence time of biochar in soil makes it an important C sequestration tool (Lehmann and others 2006). During the conversion of biomass to biochar, about 50% of the original C is retained in the biochar, which offers considerable opportunity for creating a C sink (Lehmann 2007). There is ample evidence that in certain environments, charcoal is indeed recalcitrant; however, charcoal is not a homogeneous substance (Hedges and others 2000) and certain fractions will decompose at varying rates under different conditions. It has been predicted that the stable portion of biochar has a mean residence time of greater than 1000 years (Cheng and others 2008; Lehmann and others 2008; Liang and others 2008). Deposits of charcoal up to 9500 years old have been found in wet tropical forest soils in Guyana (Hammond and others 2006), up to 6000 years old in Amazonia (Soubies 1979), and up to 23,000 years old in Costa Rica (Titiz and Sanford 2007). Bird and Grocke (1997) found that components of charred material are highly resistant to oxidation in laboratory treatments with both acid dichromate and basic peroxide, suggesting that fractions of charcoal are long-lived. Additionally, the presence of charcoal in soils and sediments from forest wildfire or prescribed burning, even after thousands of years, indicates the high persistence of black C under natural conditions (Saldarriaga and West 1986; Glaser and others 2001). Black C has been discovered in sediments that are several million years old (Herring 1985), making it up to 13,900 years older than other organic C (Masiello and Druffel 1998). Charcoal's resistance to chemical and microbiological breakdown is attributed to the polynuclear aromatic and heteroaromatic ring system structure (Haumaier and Zech 1995; Glaser and others 2002). The residence time of biochar is unknown and difficult to determine in part due to its heterogeneity. However, its stability is substantially greater than other organic matter under the same environmental conditions (Baldock and Skjemstad 2000; Liang and others 2008; Cheng and Lehmann 2009). Therefore, the transformation of labile plant organic matter into biochar through pyrolysis not only reduces CO2 emissions from energy production, but biochar additions to the soil constitute a net withdrawal of CO2 from the atmosphere.

Conclusions and Forest Management Implications

Existing research on intentional biochar application to soil has typically examined the effects of these amendments on crop yields and soil properties in agricultural systems throughout many regions. This pointed research has clarified many biochar uncertainties related to inherent biochar properties and has begun to identify mechanisms behind soil improvements. In fact, general knowledge has become so advanced that "niche" or "designer" biochars are being produced with the intention of providing ideal biochar for enhancing specific soil and crop types. This surge of inquiry, understanding, and new markets has brought about continued support and new project development to further advance the field. However, this is not the case for biochar additions to forest ecosystems. Little is known about the short- or long-term consequences of biochar application to forests, especially given that many of these fire-prone ecosystems have had analogous wildfire charcoal inputs for thousands of years. These ecosystems provide abundant and continual feedstocks in the form of residues that could be converted to biochar on-site, and forests may distinctly benefit from biochar application.

A mobile fast-pyrolysis system, when combined with forest ecosystem biochar application, offers a potential solution to excess biomass accumulation. Using the abundant forest biomass that is accumulated annually through forest harvest residues and hazard fuel reduction projects may generate biofuel that could reduce dependence on foreign and/or non-renewable energy sources. However, implementation and operational recommendations must be supported by a comprehensive, mechanistic understanding of potential site consequences to infer positive and negative effects associated with biomass removals and biochar additions across the range of site types. Demonstrating parallel results among multiple temporal and spatial-scales, such as field, laboratory, and greenhouse studies, is essential to gain a better understanding of biochar, soil, and plant interactions in soils of the Inland Northwest in association with mobile fast-pyrolysis bioenergy production systems.

The potential to improve C sequestration by adding biochar to soil creates an important opportunity to mitigate greenhouse gas emissions. While this idea is not new (Seifritz 1993), it has recently gained interest with the increasing global awareness of greenhouse gas emissions and the effects of climate change. It has even been suggested that with the use of biochar as a greenhouse gas mitigation tool, biochar C sequestration could exceed current emissions from fossil fuels, providing a net soil C sink (Lehmann and others 2006).

In forests, mobile fast-pyrolysis units could be located throughout a large region of forests. This mobility would provide opportunities to reduce hazardous forest biomass accumulations while generating biofuels and biochar, thereby creating a greater opportunity to produce C neutral biofuels and sequester C with biochar application.

The objective of forest bioenergy production systems should not be to enhance soil nutrient status and improve forest productivity with biochar additions, but instead to use the renewable and abundant forest biomass that is annually produced through forest harvest residues or hazard fuel reduction to generate biofuels, reduce wildfire risk, and improve forest health. A mobile fast-pyrolysis system offers a solution to biomass accumulation in forest ecosystems, and may improve the economic and environmental impact of biomass utilization for energy production.

REFERENCES

Agee, J.K. 1996. Fire Ecology of Pacific Northwest Forests. Island Press. 505 p.

Allen-King, R.M.; Grathwohl, P.; Ball, W.P. 2002. New modeling paradigms for the sorption of hydrophobic organic chemicals to heterogeneous carbonaceous matter in soils, sediments, and rocks. *Advances in Water Resources* 25:985-1016.

Amonette, J.E.; Joseph, S. 2009. Characteristics of biochar: Microchemical properties. In: Biochar for environmental management: Science and technology. Eds. J. Lehmann and Joseph. Earthscan, London; *Sterling*, VA: 33-52.

Asai, H.; Samson, B.K.; Stephan, H.M.; Songyikhangsuthor, K.; Homma, K.; Kiyono, Y.; Inoue, Y.; Shiraiwa, T.; Horie, 2009. Biochar amendment techniques for upland rice production in northern Laos. *Field Crops Research* 111:81-84.

Badger, P.C.; Fransham, P. 2006. Use of mobile fast pyrolysis plants to densify biomass and reduce biomass handling costs—A preliminary assessment. *Biomass and Bioenergy* 30:321-325.

Baldock, J.A.; Skjemstad, J.O. 2000. Role of the soil matrix and minerals in protecting natural organic materials against biological attack. *Organic Geochemistry* 31:697-710.

Berglund, L.M.; DeLuca, T.H.; Zackrisson, O. 2004. Activated carbon amendments to soil alters nitrification rates in Scots pine forests. *Soil Biology and Biochemistry* 36:2067-2073.

Binkley, D. 1986. Forest Nutrition Management. John Wiley & Sons: New York. 290 p.

Bird, M.I.; Grocke, D.R. 1997. Determination of the abundance and carbon isotope composition of elemental carbon in sediments. *Geochimica et Cosmochimica Acta* 61:3413-3423.

Blackwell, P.; Reithmuller, G.; Collins, M. 2009. Biochar applications to soil. In: Biochar for environmental management: Science and technology. Eds. J. Lehmann and S. Joseph. Earthscan, London; *Sterling*, VA: 207-226.

Boucher, M.E.; Chaala, A.; Roy, C. 2000. Bio-oils obtained by vacuum pyrolysis of softwood bark as a liquid fuel for gas turbines. Part 1: Properties of bio-oil and its blends with methanol and pyrolysis aqueous phase. *Biomass and Bioenergy* 19(5):337-350.

Bourke, J.; Manley-Harris, M.; Fushimi, C.; Dowali, K.; Nunoura, T.; Antal, M.J., Jr. 2007. Do all carbonized charcoals have the same chemical structure? 2. A model of the chemical structure of carbonized charcoal. *Industrial and Engineering Chemistry Research* 46:5954-5967.

Brady, N.C.; Weil, R.R. 2004. Elements of the Nature and Properties of Soils, 2nd ed. Pearson Prentice Hall: *Upper Saddle River*, NJ: 111-112.

Bridgewater, A.V., editor. 2002. Fast pyrolysis of biomass: A handbook, vol. 2. CPL Press: Newbury, United Kingdom. 26 p.

Bridgewater, A.V. 2004. Biomass fast pyrolysis. *Thermal Science* 8:21-49.

Brodowski, S.; John, B.; Flessa, H.; Amelung, W. 2006. Aggregate-occluded black carbon in soil. *European Journal of Soil Science* 57:539-546.

Bucheli, T.D.; Gustafsson, O. 2000. Quantification of the soot– water distribution coefficient of PAHs provides mechanistic basis for enhanced sorption observations. Environmental *Science and Technology* 34:5144-5151.

Bucheli, T.D.; Gustafsson, O. 2003. Soot sorption of non- ortho and ortho substituted PCBs. *Chemosphere* 53:515-522.

Caldwell. T.G.; Johnson, D.W.; Miller, W.W.; Qualls, R.G. 2002. Forest floor carbon and nitrogen losses due to prescription fire. *Soil Science Society of America Journal* 66:262-267.

Chan, K.Y.; Van Zwieten, L.; Meszaros, I.; Downie, A.; Joseph, S. 2007. Agronomic values of greenwaste biochar as a soil amendment. *Australian Journal of Soil Research* 45:629-634.

Chan, K.; Xu, Z. 2009. Biochar: Nutrient Properties and Their Enhancement. In: Biochar for Environmental Management: Science and Technology. Eds. J. Lehmann and S. Joseph. Earthscan, London; *Sterling*, VA: 53-66.

Cheng, C.H.; Lehmann,J.; Thies,J.E.; Burton, S.D.; Engelhard, M.H. 2006. Oxidation of black carbon by biotic and abiotic processes. *Organic Geochemistry* 37:1477-1488.

Cheng, C.H.; Lehmann, J.; Engelhard, M.H. 2008. Natural oxidation of black carbon in soils: Changes in molecular form and surface charge along a climosequence. *Geochimica et Cosmochimica Acta* 72:1598-1610.

Chidumayo, E.N. 1994. Effects of wood carbonization on soil and initial development of seedlings in miombo woodland, Zambia. *Forest Ecology Management* 70:353-357.

Cochrane, T.T.; Sanchez, P.A. 1980. Land resources, soil properties and their management in the Amazon region: A state of knowledge report. In: International Conference on Amazon Land Use and Agricultural Research, CIAT; Cali, Colombia.

Cohen-Ofri, I.; Popovitz-Biro, R.; Weiner, S. 2007. Structural characterization of modern and fossilized charcoal produced in natural fires as determined by using electron energy loss spectroscopy. *Chemistry—A European Journal* 13:2306-2310.

DeLuca, T.H.; Nilsson, M.C.; Zackrisson, O. 2002. Nitrogen mineralization and phenol accumulation along a fire chronosequence in northern Sweden. *Oecologia* 133:206-214.

DeLuca, T.H.; MacKenzie, M.D.; Gundale, M.J.; Holben,W.E. 2006. Wildfire-produced charcoal directly influences nitrogen cycling in forest ecosystems. *Soil Science Society of America Journal* 70:448-453.

Downie, A.; Crosky, A.; Munroe, P. 2009. Physical properties of biochar. In: Biochar for environmental management: Science and technology. Eds. J. Lehmann and S. Joseph. Earthscan, London; *Sterling*, VA: 13-32.

Evans, A.M. 2008. Synthesis of knowledge from woody biomass removal cases. Forest Guild. Santa Fe, NM.

Fisher, R.F.; Binkley, D. 2000. Ecology and Management of Forest Soils. John Wiley and Sons: New York. 489 p.

Fleming, R.L.; Powers, R.F.; Foster, N.W.; [and others]. 2006. Effects of organic matter removal, soil compaction, and vegetation control on 5-year seedling performance: A regional comparison of long-term soil productivity sites. *Canadian Journal of Forest Research* 36:529-550.

Fox, T.R.; Allen, H.L.; Albaugh, T.J.; Rubilar, R.; Carlson, C.A. 2007. Tree nutrition and forest fertilization of pine plantations in the southern United States. Southern *Journal of Applied Forestry* 31:5-11.

Garcia-Perez, M.A.; Chaala, H.; Pakdel, H.; Kretschmer, D.; Roy, C. 2007. Characterization of bio-oils in chemical families. *Biomass and Bioenergy* 31(4):222-242.

Gaskin, J.W.; Speir, A.; Morris, L.M.; Ogden, L.; Harris, K.; Lee, D.; Das, K.C. 2007. Potential for pyrolysis char to affect soil moisture and nutrient status of a loamy sand soil. In: Proceedings of the Georgia Water Resources Conference; March 27-29 2007; University of Georgia, Athens.

Gaskin, J.W.; Speir, R.A.; Harris, K.; Das, K.C.; Lee, R.D.; Morris, L.A.; Fisher, D.S. 2010. Effect of peanut hull and pine chip biochar on soil nutrients, corn nutrient status, and yield. *Agronomy Journal* 102:623-633.

Gaskin, J.W.; Steiner, C.; Harris, K.; Das, K.C.; Biben, B. 2008. Effect of low-temperature pyrolysis conditions on biochar for agricultural use. Transactions of the American *Society of Agricultural and Biological Engineers* 51:2061-2069.

Glaser, B.; Haumaier, L.; Guggenberger, G.; Zech, W. 2001. The 'terra preta' phenomenon: A model for sustainable agriculture in the humid tropics. *Naturwissenschaften* 88:37-41.

Glaser, B.; Lehmann, J.; Zech, W. 2002. Ameliorating physical and chemical properties of highly weathered soils in the tropics with charcoal—A review. *Biology and Fertility of Soils* 35:219-230.

Goldberg, E.D. 1985. Black Carbon in the Environment. John Wiley and Sons: New York.

Gundale, M.J.; DeLuca, T.H. 2006. Temperature and source material influence ecological attributes of ponderosa pine and Douglas-fir charcoal. *Forest Ecology and Management* 231:86-93.

Hammond, D.S.; Steege, H.; Van der Borg, K. 2006. Upland soil charcoal in the wet tropical forests of Central Guyana. *Biotropica* 39:153-160.

Haumaier, L.; Zech, W. 1995. Black carbon—Possible source of highly aromatic components of soil humic acids. *Organic Geochemistry* 23:191-196.

Healthy Forest Initiative. 2003. Initiative for wildfire prevention and stronger communities. The White House: Washington, DC.

Hedges, J.I.; Edlinton, G.; Gatcher, P.G.; [and others]. 2000. The molecularly-uncharacterized component of nonliving organic matter in natural environments. *Organic Geochemistry* 31:945-958.

Herring, J.R. 1985. Charcoal flux into sediments of the north Pacific Ocean: the Cenozoic record of burning. In: Eds. E.T. Sundquist and W.S. Broecker. The carbon cycle and atmospheric CO2: Natural variations from Archean to present. *American Geophysical Union Geophysical Monograph Series* 32:419-442.

Hoshi, T. 2001. Growth promotion of tea trees by putting bamboo charcoal in soil. In: Proceedings of 2001 International Conference on O-cha (Tea) Culture and Science; Tokyo, Japan: 147-150.

Hosking, J.S. 1938. The ignition at low temperatures of the organic matter in soils. *Journal of Agricultural Science* 28:393-400.

Iswaran, V.; Jauhri, K.S.; Sen, A. 1980. Effect of charcoal, coal and peat on the yield of moong, soybean and pea. *Soil Biology and Biochemistry* 12:191-192.

Jin, H.; Lehmann, J.; Thies, J.E. 2008. Soil microbial community response to amending maize soils with maize stover charcoal. In: Proceedings of the 2008 International Biochar Initiative; 8-10 September 2009; Newcastle, United Kingdom.

Johnson, D.W.; Curtis, P.S. 2001. Effects of forest management on soil C and N storage: Meta analysis. *Forest Ecology and Management* 140(2-3):227-238.

Kauffman, J.B. 1990. Ecological relationships of vegetation and fire in the Pacific Northwest. In: Tech. Coords. J.D. Walsted, S.R. Radosevich, D.B. Sandberg. Natural and prescribed fire in Pacific Northwest forests. Oregon State University Press: Corvallis, OR: 39-52.

Kimetu, J.M.; Lehmann, J.; Ngoze, S.O.; Mugendi, D.M.; Kinyangi, J.M.; Riha, S.; Verchot, L.; Recha, J.W.; Pell, A.N. 2008. Reversibility of soil productivity decline with organic matter of differing quality along a degradation gradient. *Ecosystems* 11:726-739.

Kimmins, J.P. 1997. Predicting sustainability of forest bioenergy production in the face of changing paradigms. *Biomass and Bioenergy* 13:201-212.

Kimmins, J.P. 2004. Forest ecology: A foundation for sustainable forest management and environmental ethics in forestry. Prentice Hall: Upper Saddle River, NJ: xviii, 611. 89 p.

Kishchuk, B.E.; Brockley, R.P. 2002. Sulfur availability on lodgepole pine sites in British Columbia. *Soil Science Society of America Journal* 66:1325-1333.

Kishimoto, S.; Sugiura, G. 1985. Charcoal as soil conditioner. Int. Achieve Future 5:12-23

Kleineidam, S.; Schuth, C.; Grathwohl, P. 2002. Solubility normalized combined adsorption-partitioning sorption isotherms for organic pollutants. *Environmental Science and Technology* 21:4689-4697.

Knight, H. 1966. Loss of nitrogen from the forest floor by burning. *Forestry Chronicle* 42:149-152.

Kolb, T.E.; Agee, J.K.; Fule, P.Z.; McDowell, N.G.; Pearson, K.; Sala, A.; Waring, R.H. 2007. Perpetuating old ponderosa pine. *Forest Ecology and Management* 249:141-157.

Laird, D.A. 2008. The charcoal vision: A win-win-win scenario for simultaneously producing bioenergy, permanently sequestering carbon, while improving soil and water quality. *Agronomy Journal* 100:178-181.

Laird, D.A.; Brown, R.C.; Amonette, J.E.; Lehmann, J. 2009. Review of the pyrolysis platform for coproducing bio-oil and biochar. *Biofuels Bioproducts and Biorefining* 3:547-562.

Lehmann, J. 2007. Bio-energy in the black. *Frontiers in Ecology and the Environment* 5:381-387.

Lehmann, J.; Liang, B.; Solomon, D.; Lerotic, M.; Luizão, F.; Kinyangi, F.; Schäfer, T.; Wirick, S.; Jacobsen, C. 2005. Near-edge X-ray absorption fine structure (NEXAFS) spectroscopy for mapping nano-scale distribution of organic carbon forms in soil: Application to black carbon particles. *Global Biogeochemical Cycles* 19:GB1013.

Lehmann, J.; Gaunt, J.; Rondon, M. 2006. Bio-char sequestration in terrestrial ecosystems—A review. *Mitigation and Adaptation Strategies for Global Change* 11:395-419.

Lehmann, J.; Joseph, S., Eds. 2009. Biochar for Environmental Management: *Science and Technology*. Earthscan, London. 416 p.

Lehmann, J.; Rondon, M. 2006. Bio-char soil management on highly weathered soils in the humid tropics. In: Ed. N. Uphoff. *Biological Approaches to Sustainable Soil Systems*. CRC Press: Boca Raton, FL: 517-530.

Lehmann, J.; Sohi, S. 2008. Comment on "Fire-derived charcoal causes loss of forest humus." *Science* 321:1295c.

Lehman, J., Gaunt, J. Rondon, M. 2006. Biochar sequestration in terrestrial ecosystems—A review. *Mitigation and Adaptive Strategies for Global Change* 11:403-427.

Lehmann, J.; da Silva, J.P., Jr.; Rondon, M.; Cravo, M.S.; Greenwood, J.; Nehls, T.; Steiner, C.; Glaser, B. 2002. Slash-and-char—A feasible alternative for soil fertility management in the central Amazon? In: Proceedings of the 17th World Congress of Soil Science; Bangkok, Thailand. CD-ROM. Paper no. 449:1-12.

Lehmann, J.; Solomon, D.; Kinyangi, J.; Dathe, L.; Wirick, S.; Jacobsen, C. 2008. Spatial complexity of soil organic matter forms at nanometre scales. *Nature Geoscience* 1:238-242.

Liang, B.; Lehmann, J.; Solomon, D.; Sohi, S.; Thies, J.E.; Skjemstad, J.O.; Luizão, F.J.; Engelhard, M.H.; Neves, E.G.; Wirick, S. 2008. Stability of biomass-derived black carbon in soils. *Geochimica et Cosmochimica Acta* 72:6069-6078.

Liang, B.Q.; Lehmann, J.; Sohi, S.P.; Thies, J.E.; O'Neill, B.; Trujillo, L.; Gaunt, J.; Solomon, D.; Grossman, J.; Neves, E.G.; Luizao, F.J. 2010. Black carbon affects the cycling of nonblack carbon in soil. *Organic Geochemistry* 41:206-213.

Liang, B.; Lehmann, J.; Solomon, D.; Kinyangi, J.; Grossman, J.; O'Neill, B.; Skjemstad, J.O.; Thies, J.; Luizão, F.J.; Petersen, J.; Neves, E.G. 2006. Black carbon increases cation exchange capacity in soils. *Soil Science Society of America Journal* 70:1719-1730.

Lievens, C.; Carleer, R.; Cornelissen, T.; Yperman, J. 2009. Fast pyrolysis of heavy metal contaminated willow: Influence of the plant part. *Fuel* 88:1417-1425.

Lima, I.; Marshall, W.E. 2005. Ultilization of turkey manure as granular activated carbon: Physical, chemical and adsorptive properties. *Waste Management* 25:726-32.

Lynch, D.; Mackes, K. 2003. Costs for reducing fuels in Colorado forest restoration projects. In: Conference proceedings Fire, Fuel Treatments, and Ecological Restoration. Proc. RMRS-P-29. Fort Collins, CO: U.S. Department of Agriculture, Forest Service, Rocky Mountain Research Station: 167-175.

Mahendrappa, M.K.; Pitt, C.M.; Kingston, D.G.O.; Morehouse, T. 2006. Environmental impacts of harvesting white spruce on Prince Edward Island. *Biomass and Bioenergy* 30(4):363-369.

Major, J.; Steiner, C.; Downie, A.; Lehmann, J. 2009. Biochar effects on nutrient leaching. In: Eds. J. Lehmann and S. Joseph. Biochar for environmental management: *Science and technology*. Earthscan, London: Sterling, VA: 271-287.

Masiello, C.; Druffel, E.R.M. 1998. Black carbon in deep-sea sediments. *Science* 280:1911-1913.

Mbagwu, J.S.C.; Piccolo, A. 1997. Effects of humic substances from oxidized coal on soil chemical properties and maize yield. In: Eds. J. Drozd, S.S. Gonet, N. Senesi, J. Weber. The role of humic substances in the ecosystems and in environmental protection. IHSS, Polish Society of Humic Substances: Wroclaw, Poland: 921-925.

Meurisse, R.T. 1976. Soil chemical and other properties of western hemlock soils in Oregon—their relationship to productivity. In: Eds. W.A. Atkinson and R.T. Zazoski. Western hemlock management: Proceedings of a conference; May 1976; Alderbrook, WA. Contrib. No. 34. University of Washington, College of Forest Resources, Institute of Forest Products: Seattle, WA. 317 p.

Meurisse, R.T. 1985. Properties of Andisols important to forestry. In: Eds. F.H. Beinroth, W.L. Lusio, F.P. Maldonado, H. Eswaran. Proceedings of the Sixth international soil classification workshop, Chile and Ecuador. Part 1: Papers. Washington, DC: U.S. Department of Agriculture, *Soil Management Support Services*: 53-67.

Mikan, C.J.; Abrams, M.D. 1995. Altered forest composition and soil properties of historic charcoal hearths in southeastern Pennsylvania. *Canadian Journal of Forest Research* 25:687-696.

Mohan, D.; Pittman, C.U., Jr.; Steele, P.H. 2006. Pyrolysis of wood/biomass for bio-oil: A critical review. *Energy and Fuels* 20:848-889.

Neff, J.C.; Townsend, A.R.; Gleixner, G.; Lehman, S.J.; Turnbull, J.; Bowman, W.D. 2002. Variable effects of nitrogen additions on the stability and turnover of soil carbon. *Nature* 419:915-917.

Nguyen, B.T.; Lehmann, J.; Kinyangi, J.; Smernik, R.; Riha, S.J.; Engelhard, M.H. 2008. Long-term black carbon dynamics in cultivated soil. *Biogeochemistry* 89:295-308.

Piccolo, A.; Mbagwu, J.S.C. 1990. Effects of different organic waste amendments on soil microaggregate stability and molecular sizes of humic substances. *Plant Soil* 123:27-37.

Piccolo, A.; Pietramellara, G.; Mbagwu, J.S.C. 1996. Effects of coal derived humic substances on water retention and structural stability of mediterranean soils. *Soil Use Management* 12:209-213.

Pietikäinen, J.; Kiikkilä, O.; Fritze, H. 2000. Charcoal as a habitat for microbes and its effect on the microbial community of the underlying humus. *Oikos* 89:231-242.

Preston, C.M.; Schmidt, M.W. 2006. Black (pyrogenic) carbon: A synthesis of current knowledge and uncertainties with special consideration of boreal regions. *Biogeosciences* 3:397-420.

Rackham, O. 1980. Ancient woodland: Its history, vegetation and uses in England. E. Arnold, London. 402 p.

Rondon, M.A.; Lehmann, J.; Ramirez, J.; Hurtado, M. 2007. Biological nitrogen fixation by common beans (Phaseolus vulgaris L.) increases with bio-char additions. *Biology and Fertility of Soils* 43:699-708.

Rummer, B.; Prestemon, J.; May, D.; [and others]. 2003. A strategic assessment of forest biomass and fuel reduction treatments in Western States. Washington, DC: U.S. Department of Agriculture, Forest Service. 18 p.

Saito, M.; Marumoto, T. 2002. Inoculation with arbuscilar mycorrhizal fungi: The status quo in Japan and the future prospects. *Plant and Soil* 244:273-279.

Saldarriaga, J.G.; West, D.C. 1986. Holocene fires in the northern Amazon basin. Quaternary Research (NY) 26:358-366.

Seifritz, W. 1993. Should we store carbon in charcoal? *International Journal of Hydrogen Energy* 18:405-407.

Smith, N.J.H. 1980 Anthrosols and human carrying capacity in Amazonia. *Annals of the Association of American Geographers* 70:553-566.

Sohi, S.P.; Krull, E.; Lopez-Capel, E.; Bol, R. 2010. A review of biochar and its use and function in soil. In: Advances in Agronomy, Vol 105. *Elsevier* Academic Press Inc.: San Diego, CA: 47-82.

Solomon, D.; Lehmann, J.; Kinyangi, J.; [and others]. 2007. Long-term impacts of anthropogenic perturbations on dynamics and speciation of organic carbon in tropical forest and subtropical grassland ecosystems. *Global Change Biology* 13:511-530.

Sombroek, W.G. 1966. Amazon soils. A reconnaissance of the soils of the Brazilian Amazon region. Dissertation. Wageningen, The Netherlands.

Sombroek, W.G.; Nachtergaele, F.O.; Hebel, A. 1993. Amounts, dynamics and sequestering of carbon in tropical and subtropical soils. *Ambio* 22:417-426.

Soubies, F. 1979. Existence d'une Phase seche en Amazonie bresilienne date´e par la presence de charbons de bois (6000-3000 ans B.P.). [Existence of a dry phase in the Brazilian Amazon by the presence of charcoal (6000-3000 years B.P.).] Cah. ORSTOM, Ser. Geologie 133-148.

Sullivan, D.M.; Miller, R.O. 2001. Compost quality attributes, measurements and variability. In: Eds. P.J. Stofella and B.A. Kahn. Compost utilization in horticultural cropping systems. CRC Press: Boca Raton, FL: 95-120.

Steinbeiss, S.; Gleixner, G.; Antonietti, M. 2009. Effect of biochar amendment on soil carbon balance and soil microbial activity. *Soil Biology and Biochemistry* 41:1301-1310

Steiner, C.; Teixeira, W.G.; Lehmann, J.; Nehls, T.; Macedo, J.L.V.; Blum, W.E.H.; Zech, W. 2007. Long term effects of manure, charcoal and mineral fertilization on crop production and fertility on a highly weathered central Amazonian upland soil. *Plant and Soil* 291:275-290.

Stokes, B.J.; McDonald, T.P.; Kelley, T. 1993. Transpirational drying and costs for transporting woody biomass—A prelimary review. In: Proceedings of the International Energy Association Task IX, Activity 6: Transport and Handling; May 16-251993; New Brunswick, Canada. Aberdeen University: Aberdeen, United Kingdom: 76-91.

Tiedemann, A.R. 1987. Combustion losses of sulfur from forest foliage and litter. *Forest Science* 33:216-223.

Tisdale, S.L.; Nelson, W.L. 1975. Soil fertility and fertilizers. Macmillan Publishing Co., Inc.: New York. 694 p.

Titiz, B.; Sanford, R.L., Jr. 2007. Soil charcoal in old-growth rain forests from sea level to the Continental Divide. *Biotropica* 39:673-682.

Topoliantz, S.; Pong, J.F.; Ballof, S. 2005. Manioc peel and charcoal: A potential organic amendment for sustainable soil fertility in the tropics. *Biology and Fertility of Soils* 41:15-21.

Tryon, E.H. 1948. Effect of charcoal on certain physical, chemical, and biological properties of forest soils. *Ecolological Monographs* 18:82-115.

U.S. Department of Agriculture Forest Service [USDA]. 2003. Large fire cost reduction action plan. Washington, DC: U.S. Department of Agriculture, Forest Service. 24 p.

U.S. Department of Agriculture Forest Service [USDA]. 2005. A strategic assessment of forest biomass and fuel reduction treatments in western states. Gen. Tech. Rep. RMRS-GTR-149. Fort Collins, CO: U.S. Department of Agriculture, Forest Service, Rocky Mountain Research Station. 17 p.

Van Zwieten, L.; Kimber, S.; Downie, A.; Chan, K.Y.; Cowie, A.; Wainberg, R.; Morris, S. 2007. Papermill char: Benefits to soil health and plant production. In: *Proceedings of the Conference of the international agrichar initiative*: 30 April–2 May 2007, Terrigal, NSW, Australia.

Van Zwieten, L.; Kimber, S.; Morris, S.; Chan, K.Y.; Downie, A.; Rust, J.; Joseph, S.; Cowie. S. 2010. Effects of biochar from slow pyrolysis of papermill waste on agronomic performance and soil fertility. *Plant and Soil* 327:235-246.

Van Zwieten, L.; Singh, B.; Joseph, S.; Kimber, S.; Cowie, A.; Chan, K.Y. 2009. Biochar and emissions of non-CO2 greenhouse gasses from soil. In: Eds. J. Lehmann and S. Joseph. Biochar for environmental management: Science and technology. Earthscan, London: *Sterling*, VA: 227-249.

Wardle, D.A.; Nilsson, M.C.; Zackrisson, O. 2008a. Fire-derived charcoal causes loss of forest humus. *Science* 320:629.

Wardle, D.A.; Nilsson, M.C.; Zackrisson, O. 2008b. Response to comment on "Fire-derived charcoal causes loss of forest humus." *Science* 321:1295.

Wardle, D.A.; Zackrisson, O.; Nilsson, M.C. 1998. The charcoal effect in boreal forests: Mechanisms and ecological consequences. *Oecologia* 115:419-426.

Warnock, D.D.; Lehmann, J.; Kuyper, T.W.; Rillig, M.C. 2007. Mycorrhizal responses to biochar in soil—concepts and mechanisms. *Plant and Soil* 300:9-20.

Winsley, P. 2007. Biochar and bioenergy production for climate change mitigation. *New Zealand Scientific Review* 64:5-10.

Woods, W.I.; McCann, J.M. 1999. The anthropogenic origin and persistence of Amazonian dark earths. *Yearbook of the Conference of Latin Americanist Geographers* 25:7-14.

Yamato, M.; Okimori, Y.; Wibowo, I.F.; Anshori, S.; Ogawa, M. 2006. Effects of the application of charred bark of Acacia mangium on the yield of maize, cowpea and peanut, and soil chemical properties in South Sumatra, Indonesia. *Soil Science and Plant Nutrition* 52:489-495.

Yu, X.Y.; Ying, G.G.; Kookana, R.S. 2006. Sorption and desorption behavior of diuron in soil amended with charcoal. *Journal of Agricultural and Food Chemistry* 54:8545-8550.

Zhurinsh, A.; Zandersons, J.; Dobele, G. 2005. Slow pyrolysis studies for utilization of impregnated waste timber materials. *Journal of Analytical and Applied Pyrolysis* 74:439-444.

In: Carbon Considerations
Editors: W. Green and L. Wayman ISBN 978-1-62618-435-0

Chapter 5

BIOMASS FEEDSTOCKS FOR BIOPOWER: BACKGROUND AND SELECTED ISSUES*

Kelsi Bracmort

SUMMARY

Biopower—a form of renewable energy—is the generation of electric power from biomass feedstocks. Biopower, which comprised about 1% of electricity generation in 2008, may reduce greenhouse gas emissions, provide energy security, and promote economic development. A large range of feedstocks can be used, from woody and herbaceous biomass to agricultural residues. Each feedstock has technical and economic advantages and challenges compared to fossil fuels.

Unlike wind or solar energy, a biopower plant is considered to be a baseload power source because some biomass feedstocks can be used for continuous power production. However, ensuring a sustainable supply of biomass feedstocks is a major challenge. Although there are multiple biopower technologies, few of them except combustion have been deployed at commercial scale nationwide.

Federal policymakers are supporting biopower through feedstock supply analysis and biopower technology assessments. However, there is limited comprehensive data about the type and amount of biomass feedstock available to meet U.S. biopower needs at a national level. If the

* This is an edited, reformatted and augmented version of a Congressional Research Service publication, CRS Report for Congress R41440, from www.crs.gov, dated January 20, 2011.

use of dedicated biomass feedstocks to generate biopower were to develop into a sizeable industry, concerns would likely include the effect of the industry on land use (i.e., how much land would it take to grow the crops needed to fuel or co-fuel power plants) and the effect on the broader economy, including farm income and food prices. To date, these have not been issues: most existing biomass feedstocks have been waste products generated by the forest products industry or by farms, or municipal solid waste for which combustion served as both a disposal method and a source of energy.

Growing crops for use as a power source would be different from using waste. Under generally accepted assumptions regarding crop yields and energy content, approximately 31 million acres— roughly the amount of land in farms in Iowa—would be needed to supply enough biomass feedstock to satisfy 6% of total 2008 U.S. electricity retail sales. When added to the amount of land needed to meet the requirements of the Renewable Fuel Standard (RFS), a federally mandated transportation fuel requirement, the potential impacts could be significant: the RFS already consumes 35% of the nation's corn crop, and its requirements will triple between 2010 and 2022 (although much of this fuel will come from feedstocks other than corn).

Beyond land use and economic impacts, others are concerned that the use of biomass feedstocks to generate biopower, particularly through combustion, could add to greenhouse gas (GHG) emission levels and exacerbate climate change concerns. They fear that certain areas may be unsustainably harvested to meet biomass feedstock demand, or that less biomass may be left for other purposes (e.g., wood and paper products). The concerns exist partly because biomass used for biopower does not face the same constraints as biomass used for liquid transportation fuels under the RFS. In addition, the idea that biomass combustion is carbon-neutral is under scrutiny. The Environmental Protection Agency has not exempted biomass combustion emissions from the Prevention of Significant Deterioration and Title V Greenhouse Gas Tailoring Rule, although it announced plans to defer for three years GHG permitting requirements for carbon dioxide emissions from biomass-fired and other biogenic sources (those produced by living organisms). In the PSD and Title V Permitting Guidance for Greenhouse Gases issued in 2010, EPA noted that biomass could be considered a best available control technology (BACT)—a pollution control standard mandated by the Clean Air Act.

INTRODUCTION

The production of bioenergy—renewable energy derived from biomass—could potentially increase national energy security, reduce greenhouse gas emissions, and contribute to rural economic growth. Legislative, research, and industrial attention have focused on the production of bioenergy in the form of liquid transportation fuels (e.g., corn-based ethanol).[1] Biopower—the production of electricity from biomass feedstocks—may require new national policies or incentives if Congress decides to encourage its development.

Biopower, or biomass power, comprised about 1% of electricity generation in 2008.[2] It was the third-largest renewable energy source for electricity generation in that year, after conventional hydroelectric power and wind.[3] The Department of Energy's (DOE's) Energy Information Administration (EIA) projects that electricity generation from biomass will grow from 0.9% of total generation in 2008 to 5.5% in 2035.[4] The DOE reference case for this projection assumes extension of federal tax credits, state requirements for renewable electricity generation, and the loan guarantee program in the Energy Policy Act of 2005 (EPAct05; P.L. 109-58) and the American Recovery and Reinvestment Act of 2009 (ARRA; P.L. 111-5).

Current concerns for accelerating biopower growth include the need for a continuously available feedstock supply, a commercial-scale facility to generate the biopower, and market certainty for investors and purchasers alike. Improved feedstock availability, technological advancements, and new forms of economic support could increase the relative contribution of biopower to meeting U.S. energy demand.

One reason for the projected growth of biopower is the fuel's ability to be used in a baseload power plant. Baseload power is the minimum amount of electric power delivered or required over a given period of time at a steady rate. If a plant operates as a baseload plant, the plant can run continually except for maintenance and outages. With sufficient feedstock supplies, a biopower plant could provide "firm" power for baseload needs (and long-term contracts would reduce risk). In contrast, wind and solar energy require either a form of power storage, such as batteries, or a backup power source, such as natural gas turbines, in order to provide firm power.

Power generation from biomass is not limited to a specific feedstock and therefore is relatively flexible in terms of fuel suppliers. Each region of the country can pursue biomass feedstocks that are native and readily available (e.g., corn stover in the Midwest, hybrid poplar in the Northwest, switchgrass in the Southeast). The economic climate for biopower dictates that biopower plants should be located in close proximity to feedstocks to reduce transportation costs, which can be significant.[5] Furthermore, existing combustion plants can be retrofitted for biopower production; power from these plants could use existing transmission infrastructure. Financing and siting of new transmission infrastructure could add uncertainty to a proposed project.

The availability and cost of biomass feedstocks determine the amount of biopower that can be produced nationally. An overarching concern is maintaining a sustainable biomass feedstock supply.[6] If feedstocks are collected without regard to replenishment, or in an otherwise unsustainable manner, biopower enterprises may lead to natural resource deterioration such as soil erosion or the depletion of forested land. The Renewable Fuel Standard (RFS), expanded under the Energy Independence and Security Act of 2007 (EISA; P.L. 110-140), mandates a minimum volume of biofuels to be used in the national transportation fuel supply each year. Under the RFS, biomass used for renewable fuel for transportation purposes cannot be removed from federal lands, and the law excludes crops from forested lands.[7] Thus far, biomass used for biopower is not subject to the same constraints as biomass used for liquid transportation fuels under the RFS. Additionally, feedstock diversity is a formidable challenge to biopower growth, because cultivation, harvest, storage, and transport vary according to the feedstock type. Another challenge is accounting for the amount of feedstock available for biopower production due to market fluctuations and weather variability.

In considering congressional action to broaden legislative authorities for sustainable biopower production, an understanding of the various biomass feedstocks and challenges to biopower production could be useful to policymakers. This report provides analyses of commonly discussed biomass feedstocks and their relative potential for power generation. Additional biopower issues—feedstock accessibility, the biomass power plant carbon-neutrality debate, and unintended consequences of legislative activities to promote bioenergy—are also discussed.

What Kind of Biomass Is Available for Biopower?

The type, amount, and costs of biomass feedstocks available for biopower will largely determine whether biopower can thrive as a major renewable energy alternative. There is limited comprehensive data on the amount of biomass feedstocks available to meet current and future biopower needs at a national level. The supply data available is generally evaluated in terms of meeting biofuel demand. Some may argue that feedstock assessments for biofuels are adequate for biopower purposes, as the same feedstock may be used to meet both biofuel and biopower demands. Information that identifies which feedstocks exhibit the most potential for power generation in the near and long term is also scarce.[8] Furthermore, ideal or feasible locations where feedstocks may be grown are not well assessed. The potential inclusion of genetically modified dedicated energy crops or selective breeding for bioenergy purposes may alter the amount of biomass feedstock available for biopower production (and may alter land use).

Additional legislative action concerning financial support of biopower may depend on better data to estimate the economic viability of biopower plants nationwide.[9] Costs associated with biomass storage and transportation to a biopower plant, as well as other economic and environmental considerations, are among the factors assessed in individual biopower plant feasibility studies. These factors are key to determining which biomass feedstocks can be used.

In addition to economics, biological characteristics play a large role in determining the suitability of any type of biomass. Biomass is organic matter that can be converted into energy. Plants use photosynthesis to store energy (carbon-based molecules) within cell walls, and that energy is released when the biomass undergoes a biological process such as anaerobic digestion, or a chemical process such as combustion. Biomass can include land- and water-based vegetation (e.g., trees, algae), as well as other organic wastes (see Table 1).

Several types of feedstocks can be used as a fuel source for electric power generation. Primary biomass feedstocks are materials harvested or collected directly where they are grown (e.g., grains). Secondary biomass feedstocks are by-products of the processing of primary feedstocks (e.g., corn stover). Tertiary biomass feedstocks include post-consumer residues and wastes (e.g., construction and demolition waste). Appendix A shows the energy value, crop yield, advantages, disadvantages, and general comments for selected biomass feedstocks and fossil fuels for comparison.

Biomass would have to be grown in enormous quantities if it is to be used as a power source to satisfy a significant portion of national energy demand. For example, approximately 31 million acres—roughly the amount of land in farms in Iowa—of managed crops with a yield of 6 dry tons per acre per year would be needed to supply enough biomass feedstock to satisfy 6% of total 2008 U.S. electricity retail sales.[10] Quintessential biomass crops grown specifically for energy generation (i.e., dedicated energy crops) are being considered to meet energy demand. Dedicated energy crops may possess several desirable characteristics: high yield, low energy input to produce, low cost, low nutrient requirements, low contaminant level, pest resistance, and low fertilizer input.[11]

Table 1. General Classification of Biomass

Biomass groups	Biomass sub-groups, varieties and species
Wood and woody biomass	Coniferous or deciduous (gymnosperm or angiosperm); stems, branches, foliage, bark, chips, lumps, pellets, briquettes, sawdust, sawmill and other wastes from various woody species
Herbaceous and agricultural biomass	Annual or perennial and field-based or process-based such as: —grasses and flowers (alfalfa, arundo, bamboo, bana, brassica, cane, miscanthus, switchgrass, timothy, others); —straws (barley, bean, flax, corn, mint, oat, rape, rice, rye, sesame, sunflower, wheat, others); —other residues (fruits, shells, husks, hulls, pits, pips, grains, seeds, coir, stalks, cobs, kernels, bagasse, food, fodder, pulps, cakes, others)
Aquatic biomass	Marine or freshwater algae and microalgae; macroalgae (blue, green, blue-green, brown, red); seaweed, kelp, lake weed, water hyacinth, others
Animal and human biomass wastes	Bones, meat-bone meal, chicken litter, various manures, others
Contaminated biomass and industrial biomass	Municipal solid waste, demolition wood, refuse-derived fuel,
wastes (semi-biomass)	sewage sludge, hospital waste, paper-pulp sludge and liquors, waste papers, paperboard waste, chipboard, fibreboard, plywood, wood pallets and boxes, railway sleepers, tannery waste, others
Biomass mixtures	Blends from the above varieties

Source: Stanislav V. Vassilev, David Baxter, and Lars K. Andersen, et al., "An Overview of the Chemical Composition of Biomass," *Fuel*, vol. 89 (2010), pp. 913-933. Adapted by CRS.

From Biomass to Biopower

Biomass can be converted to biopower via thermo-chemical and bio-chemical conversion processes. These processes include combustion (or firing), pyrolysis, gasification, and anaerobic digestion (see box, below, and Figure 1). The technologies are at varying stages of maturity (see Figure 2).

The choice of conversion technique selected for a specific biomass feedstock results in differing amounts of useful energy recovered and forms for that energy.[12] The systems can range substantially in scale. Small-scale systems (or modular units) may be an optimal choice for rural areas with limited electricity demand. Large-scale systems may be more economically suitable in more urbanized areas or near grid connections if feedstocks are ample.

The volume of biomass feedstock supply necessary to run a biopower plant depends on the feedstock's energy content—the less the energy value, the more volume is needed. The growing area needed to produce the biomass that will supply a biopower plant is contingent not only on the energy value of the feedstock, but also on the power plant capacity, the power plant efficiency, and the feedstock yield (see Table 2). Power plant capacity is the maximum output of power, commonly expressed in millions of watts (megawatts, MW), that generating equipment can supply over a certain time period. Power plant efficiency is the amount of electric energy produced per unit of feedstock input. In general, the higher the yield of the biomass feedstock, the less growing area is required to produce a MW of power. Also, less biomass is needed to support power plants with high efficiency rates.

Selected Biopower Conversion Processes Defined

Combustion is the burning of biomass in a power plant. The biomass is burned to heat a boiler and create steam. The steam powers a turbine, which is connected to a generator to produce electricity. Existing plant efficiencies are in the low 20% range, although methods are available to advance efficiency to upwards of 40%. ("Efficiency" describes which percentage of the feedstock processed is actually converted to electricity.) Approximately 180 combustion units for biomass are in operation using wood and agricultural residues as the feedstock.

Co-firing, the simultaneous firing of biomass with coal in an existing power plant, is the most cost-effective biopower technology. Co-firing with biomass using existing equipment is less expensive than constructing a new biopower plant. The existing plant does require retrofitting to accept the biomass entering the plant. Certain air particulates associated with coal combustion are reduced with co-firing, as less coal is being burned. Co-firing has a generation efficiency in the 33%-37% range; coal-fired plants have efficiencies in the 33%-45% range. Approximately 78 co-firing units for biomass are in operation using wood and agricultural residues as the feedstock.

Gasification is the heating of biomass into synthesis gas (syngas, a mixture of hydrogen and carbon monoxide) in an environment with limited oxygen. The flammable syngas can be used in a combined gas and steam turbine to generate electricity. Generation efficiencies range from 40% to 50%. One challenge for gasification is feedstock logistics (e.g., cost to ship or transport the feedstock to the power plant). A wide variety of feedstocks could undergo gasification, including wood chips, sawdust, bark, agricultural residues, and waste. There are currently no gasification systems for biomass at any scale.

Pyrolysis is the chemical breakdown of a substance under extremely high temperatures (400°C -500°C) in the absence of oxygen. There are fast and slow pyrolysis technologies. Fast pyrolysis technologies could be used to generate electricity. Fast pyrolysis of biomass produces a liquid product, pyrolysis oil or bio-oil, that can be readily stored and transported. The bio-oils produced from these technologies would be suitable for use in boilers for electricity generation. One of the challenges with pyrolysis is that the bio-oil produced tends to be low-quality relative to what is needed for power production. Commonly used feedstock types for pyrolysis include a variety of wood and agricultural resources.

There are currently no commercial-scale pyrolysis facilities for biomass.

Anaerobic digestion (not shown in Figure 1) is a biological conversion process that breaks down a feedstock (e.g., manure, landfill waste) in the absence of oxygen to produce methane, among other outputs, that can be captured and used as an energy source to generate electricity. Anaerobic digestion systems have historically been used for comparatively smaller-scale energy generation in rural areas. Feedstocks suitable for digestion include brewery waste, cheese whey, manure, grass clippings, restaurant wastes, and the organic fraction of municipal solid waste, among others. Generation efficiency is roughly 20%-30%. Approximately 150 anaerobic digesters are in operation using manure as the feedstock.

Sources: Oak Ridge National Laboratory, *Biomass Energy Data Book: Edition 2*, ORNL/Tm-2009/098, December 2009, http://cta.ornl.gov/bedb/pdf/BEDB2_Full_Doc.pdf. International Energy Agency, *Biomass for Power Generation and CHP*, ETE03, January 2007, http://www.iea.org/techno/essentials3.pdf. National Association of State Foresters, *A Strategy for Increasing the Use of Woody Biomass for Energy*, Portland, ME, September

2008, http://www.stateforesters.org/files/NASF-biomass-strategy-FULL-REPORT-2009.pdf. Sally Brown, "Putting the Landfill Energy Myth to Rest," *BioCycle*, May 2010. John Balsam and Dave Ryan, *Anaerobic Digestion of Animal Wastes: Factors to Consider*, ATTRA—National Sustainable Agriculture Information Service, IP219, 2006, http://attra.ncat.org/attra-pub/anaerobic.html. Jennifer Beddoes, Kelsi Bracmort, and Robert Burns et al., *An Analysis of Energy Production Costs from Anaerobic Digestion Systems on U.S. Livestock Production Facilities*, USDA Natural Resources Conservation Service, October 2007. Personal communication with Robert Baldwin, National Renewable Energy Laboratory, 2010. Personal communication with Lynn Wright, biomass consultant working with Oak Ridge National Laboratory. For more information on anaerobic digestion, see CRS Report R40667, *Anaerobic Digestion: Greenhouse Gas Emission Reduction and Energy Generation*, by Kelsi Bracmort.

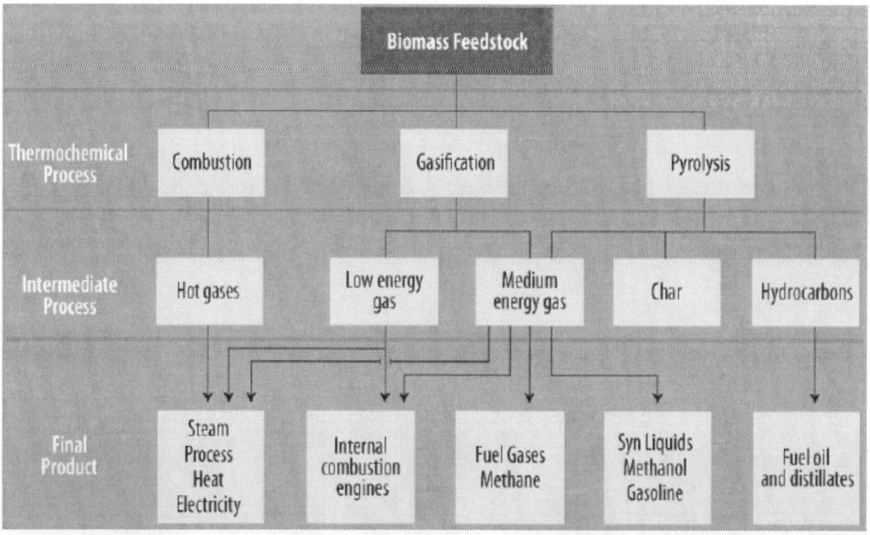

Source: Peter McKendry, "Energy Production from Biomass (Part 2): Conversion Technologies," *Bioresource Technology*, vol. 83 (2002), pp. 47-54. Adapted by CRS.

Figure 1. Biopower Conversion Processes.

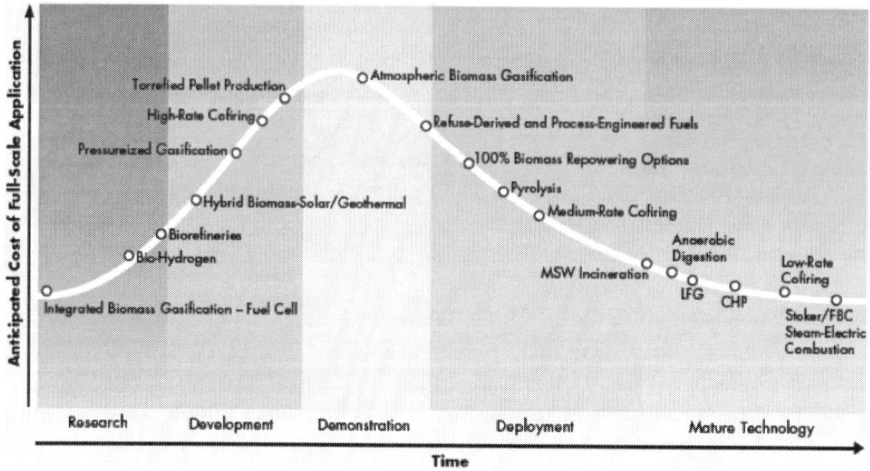

Source: Electric Power Research Institute, Biopower Generation: Biomass Issues, Fuels, Technologies, and Research, Development, Demonstration, and Deployment Opportunities, February 2010.

Figure 2. Biopower and Biofuel Technology Pipeline.

Table 2. Biomass Feedstock Growing Area Required to Produce Biopower

Power plant capacity factor (%)	80	80	80	80	90	90	90	90
Power plant efficiency (%)	25	30	35	40	25	30	35	40
Crop yield (dry tons/acre/year)								
(see Appendix A)			Growing Acres Required per MW					
1	5930	4941	4235	3706	6671	5559	4765	4169
2	2965	2471	2118	1853	3335	2780	2382	2085
3	1977	1647	1412	1235	2224	1853	1588	1390
4	1482	1235	1059	927	1668	1390	1191	1042
5	1186	988	847	741	1334	1112	953	834
6	988	824	706	618	1112	927	794	695
7	847	706	605	529	953	794	681	596
8	741	618	529	463	834	695	596	521
9	659	549	471	412	741	618	529	463
10	593	494	424	371	667	556	476	417

Source: Department of Energy, Relationship Between Power Plant Efficiency and Capacity and Tons Biomass Required and Acres Required, Lynn Wright, http://bioenergy

Notes (from original source): Raw numbers have been used in the above table. Calculations assume dry biomass at 8500 btu/lb = 19.75 Gj/MG and 3413 btu/kWH = 0.0036 Gj/kWh.

Rule of thumb relationship of 1000 acres and 5000 dry tons per MW is based on 80% capacity, 30% efficiency, and 5 dry ton/acre/year yield. A program goal would be to have a relationship of 500 acres and 4200 dry tons per MW at 90% capacity, 40% efficiency, and 8 dry ton/acre/year yield.

Yields of 1-2 dry ton/acre/year are common for natural forests but could also represent residue levels available from high yield plantations. Yields of 3-4 dry ton/acre/year are common for pulpwood pine plantations. Yields of 4-7 dry ton/acre/year are being observed in woody crop and herbaceous crop plantings without irrigation,

5dt/ac/yr still best average estimate. Yields of 7-10 dry ton/acre/year are being observed in some energy crop plantings with best clones or varieties and/or with irrigation or high water tables.

Total planted area or growing area required to supply a biomass facility should be used rather than area actually being harvested in any given year. While these are the same for a herbaceous crop harvested annually, they differ

significantly for a woody crop harvested once every few years. Calculation of the annual harvested area for a wood crop requires knowing both the yield (dry ton/acre/year) and the harvest age of the woody crop. This varies from project to project.

CARBON BALANCE

Certain sources of biomass (e.g., forestry products, dedicated energy crops) are deemed by some to be carbon-neutral because they absorb enough CO2 during their growth period to balance the release of CO2 when they are burned for energy (see Figure 3). The term *carbon-neutral* is generally defined as the combustion or oxidation of matter which causes no net increase in GHG emissions on a lifecycle basis.[13] One controversial aspect of the carbon neutrality debate, and what requires further study, is the magnitude in which these plant-derived feedstocks will be used for energy production and thus whether the feedstock supply can be sustained (or replenished) without environmental impairment. Some examples of environmental impairment involve disrupting forest ecosystems by cutting down large amounts of trees, or affecting the climate by not capturing GHGs emitted during bioenergy production. If the feedstocks are not replenished so that they can absorb $_{CO2,}$ or GHG emissions are not captured from a biopower plant, the resulting GHG releases can be akin to that of carbon-positive fossil fuels.

Source: John A. Matthews, "Carbon-Negative Biofuels," *Energy Policy*, vol. 36 (2008), pp. 940-945; Biopact, "The Strange World of Carbon-Negative Bioenergy: The More You Drive Your Car, the More You Tackle Climate Change," 2007, http://news.mongabay.com/bioenergy. Adapted by CRS.

Notes: Carbon-positive fuels are burned, releasing CO_2 into the atmosphere. Carbon-neutral fuels absorb CO_2 as they grow and release the same carbon back into the atmosphere when burnt. Carbon-negative fuels absorb CO_2 as they grow and release less than this amount into the atmosphere when used as fuel, either through directing part of the biomass as biochar back into the soil or through carbon capture and sequestration.

Figure 3. Carbon Balance of Energy.

The premise that biomass energy is carbon neutral is under scrutiny.[14] Some argue that biomass combustion could increase GHG emissions if biomass feedstock is handled in an unsustainable manner, among other concerns. Others contend that biopower plant emissions add no new carbon to the atmosphere because biopower plants use only materials that would decay (e.g., residuals, byproducts, and thinnings). Under the final Prevention of Significant Deterioration (PSD) and Title V Greenhouse Gas Tailoring Rule, PSD requirements apply to projects that increase net GHG emissions by at least 75,000 tons per year CO2 equivalent.[15] The rule grants exemptions based on tonnage amounts, not emissions based on source category (e.g., biomass combustion). The Tailoring Rule does not exempt emissions from biomass combustion.[16] EPA has decided to defer for three years permitting requirements for CO2 emissions from biomass-fired and other biogenic sources.[17] Moreover, EPA decided that certain types of biomass could be regarded as a best available control technology (BACT).[18] Further guidance will be issued by EPA in early 2011 on how to include biomass in the BACT selection process. PSD permits require that facilities apply the BACT. BACT is determined by individual states with EPA guidance on a case-by-case basis.[19]

There are other aspects associated with the designation of biomass energy as carbon neutral, many of which are beyond the scope of this report.

IMPLICATIONS FOR LEGISLATION

Biopower straddles at least three legislative areas: agriculture, energy, and environment. The main benefits that agricultural legislation could provide, as argued by proponents for biopower, are to ensure an adequate feedstock supply, maintain productive field conditions during biomass growth and harvest, and assist farmers who participate in the bioenergy market. Energy objectives, as stated by supporters, involve establishing a robust biopower technology platform and providing financial and technical assistance for biopower technology pioneers. Protecting the environment throughout the biomass-to-biopower conversion is the major environmental objective, including monitoring GHG emissions released during energy production.

As a candidate for large-scale energy use, the biopower industry may challenge Congress to address its evolving needs on a frequent basis until biopower is a seasoned energy alternative. One frequent topic of discussion for renewable energy is the "uneven" playing field for certain feedstocks. Supporters of pre-selected feedstocks for biopower production argue that resources can be targeted to that handful of feedstocks that display the most potential for bioenergy production. Opponents contend that pre-selecting certain feedstocks makes it difficult for other feedstocks to obtain the support needed to show their competitiveness as a biopower source.

Congress currently supports biopower with the Renewable Energy Production Tax Credit (PTC) and the Investment Tax Credit (ITC). The PTC is an incentive to business developers of renewable energy projects producing electricity, whereby a developer can apply for a credit against taxes for each kilowatt-hour of renewable energy produced.[20] The ITC is an incentive for domestic investment in renewable energy plants and equipment.[21] In lieu of tax credits, biopower administrators may apply for payments from the Department of the Treasury under the Section 1603 Program for having placed in service a specified energy property.[22] Qualifying projects are those that are placed in service or under construction before the end of 2011. Moving forward, there may be unintended consequences of legislation that supports biopower. For example, initial USDA regulations for implementing the Biomass Crop Assistance Program (BCAP) led to shifting sawmill residues from products (especially particleboard) to energy rather than increasing utilization of forest waste or planting biomass feedstocks for bioenergy.[23]

Legislative efforts are under way to further support the biopower industry. One relevant legislative effort is the creation of a renewable electricity standard (RES) to encourage renewable energy use, and thus production of renewable energy such as biopower. One energy policy bill, introduced during the 111th Congress, that included a federal RES was the American Clean Energy Leadership Act of 2009 (ACELA, S. 1462).[24] The RES would have required utilities that sell electricity to consumers to obtain a percentage of their annual electricity supply from renewable energy sources or energy efficiency, starting at 3% in 2011 and rising incrementally to 15% by 2021. S. 1462 identified biomass as an eligible renewable source. H.R. 2454, H.R. 890, S. 433, and S. 3021 were other bills introduced during the 111th Congress that would have created a federal RES. H.R. 2454, the American Clean Energy and Security Act, contained provisions that would have supported biopower, such as transmission planning and net metering, along with an RES.[25]

Conclusion

While there remain significant challenges to its future development, biopower production could increase in the coming years to satisfy U.S. renewable energy demand (e.g., state renewable portfolio standards). Generation of electricity from biopower plants has advantages over other renewable sources such as wind and solar. Biopower plants are considered baseload plants. Also, multiple biomass feedstocks can be used to generate electricity. Some disadvantages of using biomass for electricity generation include the cost to transport the biomass to the biopower plant, less biomass to be used for other purposes, and environmental tensions such as whether biomass combustion is carbon-neutral. A sustainable supply of biomass feedstocks could be favorable to biopower growth.

Appendix A. Biomass Feedstock Characteristics for Biopower Generation

Feedstock Type	Energy Value Btu/lb (dry)[a]	Feedstock Yield[b]	Selected Advantages	Selected Disadvantages	Comments[c]
Woody Biomass					
Willow (example of a wood crop grown as a bush type or "coppice" crop in high density plantings as dedicated bioenergy crop)	7,983-8,497	4-8 dry tons/acre/year harvested on 2-4 year cycle	• High yield potential • Grown for several cycles before replanting • Select varieties easily replicated by cloning • Easy to automate planting and harvest as a row crop • Short harvest cycle for wood • Farmers can grow and harvest • Low ash content	• Requires specialized harvesting equipment • High density plantings are costly to establish • U.S. experience and varieties of willow currently limited to Northeast • Must be harvested in winter to obtain regrowth for several cycles • Agricultural site preparation needed for successful establishment • Susceptibility of some willow varieties to insects and diseases may require occasional chemical applications	• Very high future yield potential with genetic selection • Innovative harvest equipment is available • Many woody hardwood crops can be grown as bush type crops • Economic yields obtained on marginal to good cropland • Less fertilization required than agricultural crops

Appendix A. (Continued).

Feedstock Type	Energy Value Btu/lb (dry)[a]	Feedstock Yield[b]	Selected Advantages	Selected Disadvantages	Comments[c]
Hybrid poplar (example of a fast growing hardwood grown as a row crop for bioenergy or multiple purposes)	8,183-8,491	3-7 dry tons/acre/year; harvested on 5-15 year cycles	• High yield potential • Select varieties easily replicated by cloning • Easy to automate planting and harvest as a row crop • Can be stored on stump until needed • Relatively low-maintenance crop • Improvements for bioenergy will also likely benefit the pulp and paper industry	• No immediate return on investment • Susceptibility of some hybrid poplar varieties to insects and diseases may require occasional chemical applications • Agriculture-type site preparation needed for successful establishment • Regrowth after harvest is possible but replanting with superior clones is recommend	• Very high future yield potential with genetic selection • Innovative harvest equipment is under development • Economic yields obtained on marginal to good cropland
Loblolly pine (example of fast-growing softwood grown as a row crop for bioenergy or multiple purposes)	8,000-9,120	3-7 dry tons/acre/year; harvested every 20-40 years	• 30 million acres of southern pines already are being managed in southern U.S. • Somewhat higher energy value than poplars and willows	• Pines cannot currently be cloned; standard breeding and family selection techniques must be used to improve yield	• Well suited for thermal technologies to generate electricity and ethanol • Conversion to liquid fuels is possible with acid hydrolysis and as a co-product of pulp fiber production

Feedstock Type	Energy Value Btu/lb (dry)a	Feedstock Yieldb	Selected Advantages	Selected Disadvantages	Commentsc
			• Grows better than poplars and other hardwoods on marginal coastal plains and flatwoods soils • Valuable to landowners as a low-intensity crop with multiple markets	• Pines are mostly hand planted, since planted as rooted seedlings; Limited automation is possible • Agricultural type site preparation needed for rapid early growth	• Less fertilization required than for agricultural crops
Pine chips (example of forest residues from timber and fiber harvests)	8,000-9,120	10-20 dry tons/acre of on-site residues following logging; harvested every 20-40 years	• Relatively inexpensive if chips produced at the roadside as a by-product of wood processing • Infrastructure to handle forest residues exists	• High retrieval cost when tops and branches collected in forest due to labor-intensive collection and transportation • Tops and branches may not be accessible or environmentally sustainable to remove for chipping, depending on location and soil type	• An expanded ethanol industry using wood can also be an additional source of biopower as a co-product
Mill residue (from both sawmills and pulp mills)	7,000-10,000	Highly variable depending on operating size of the mill	• Easily available and accessible • Inexpensive • Infrastructure to handle feedstock exists	• Nearly all mill residues are currently being used in wood products such as particleboard and paper, as fuel for heat or biopower, or to make mulch	• Most mill residues will continue to be used at or near the site where wood is processed though at higher energy costs, more might shift to on-site bioenergy production

Appendix A. (Continued).

Feedstock Type	Energy Value Btu/lb (dry)[a]	Feedstock Yield[b]	Selected Advantages	Selected Disadvantages	Comments[c]
Herbaceous Biomass					
Miscanthus (highly productive grass in Europe) Switchgrass (example of several possible perennial warm-season grasses)	7,781-8,417 7,754-8,233	4-7 dry tons/acre/year current U.S. average 4-12 dry tons/acre/year has been observed for delayed harvest yields in Europe 4-9 dry tons/acre/year range in research trials	• Once established, can be harvested annually for 15-20 years before having to replant • Low fertilizer requirements • Drought-tolerant • Very high yield potential with adequate water • Long growth season in mid-U.S. • Giant miscanthus is sterile, thus not invasive • Suitable for growth on marginal land • Relatively high, reliable productivity across a wide geographical range • Low water and nutrient requirements	• No immediate harvest; takes one to three years to be established • Not a native species • Testing as a bioenergy feedstock limited to the last 10 years (most research conducted in Europe) • Thick-stem and moisture content of 30 to 50% in late fall requires specialized harvesting equipment • Planting of rhizomes requires specialized equipment • No immediate harvest; takes two to three years to be established	• Perennial grass • Established vegetatively by planting divided rhizome pieces • Higher yields are likely to occur on well-drained soils suitable for annual row crops • Suitable for thermochemical conversion processes, such as combustion, if harvest is delayed until late winter • Native perennial grass

Feedstock Type	Energy Value Btu/lb (dry)a	Feedstock Yieldb	Selected Advantages	Selected Disadvantages	Commentsc
			• Provides wildlife cover and erosion control • Can be grown and harvested with existing farm equipment • Planted by seeding • Low moisture content if harvested in late fall (15% to 20%) • Few major insect or disease pests	• May require annual fertilization to optimize yields, but at relatively low levels • Annual harvest must occur over a relatively short window of time each fall • Year-round storage is needed if switchgrass is only feedstock for a bioenergy facility • Energy content diminishes over year if not kept dry • Ash content can be high	• Can be used for gasification, combustion or pyrolysis technologies to generate electricity or for biochemical conversion to ethanol • Research for bioenergy feedstock began in the 1980s
Sorghum—varieties selected for biomass production (similar to a tall thin stalked forage sorghum crop)	7,476-8,184	4-10 dry tons/acre/year Higher yields observed	• Suitable for warm and dry growing regions • Seed production delayed, thus produces more biomass • Annual crop, thus immediate return on investment	• Yields more variable than switchgrass, with rainfall differences • Requires > 20 inches of rainfall annually • Annual crop, thus more expense and work to replant each year	• Sweet, grain, and silage sorghums are more suitable for ethanol production with higher sugar content • Susceptibility to anthracnose disease of some genotypes

Appendix A. (Continued).

Feedstock Type	Energy Value Btu/lb (dry)[a]	Feedstock Yield[b]	Selected Advantages	Selected Disadvantages	Comments[c]
			• Grows across most of eastern and central U.S., not frost limited		
Sugarcane/Energycane	7,450-8,349	Yields exceeding 10 dry tons/acre common	• Takes approximately one year to become established • Has very high yield potential in tropical, semi-tropical and subtropical regions of world • A multi-purpose crop-producing sugar (or ethanol) and biopower feedstock • Drought-adapted	• Planting locations limited to a few states in the South and Hawaii • Must be replanted every 4 to 5 years • Planting is vegetative (stalks are laid down) rather than by seed • Vulnerable to bacterial, fungal, viral, and insect pests • Crop must be harvested green and dewatered or stored like silage	• Literature mostly centers on its use for ethanol • The bagasse (residue once juice is extracted from the sugarcane) may be used for biopower e.g., frequently used in Brazil • Research ongoing to hybridize to achieve cold tolerance

Feedstock Type	Energy Value Btu/lb (dry)[a]	Feedstock Yield[b]	Selected Advantages	Selected Disadvantages	Comments[c]
Aquatic Biomass					
Algae	8,000-10,000 for algal mass; 16,000 for algal oil and lipids	Estimates not available for biopower	• Cultivation strategies can minimize or avoid competition with arable land and nutrients used for conventional agriculture • Can use waste water, produced water, and saline water, reducing competition for limited freshwater supplies • Can recycle carbon from CO2-rich flue emissions from stationary sources including power plants and other industrial emitters	• Relatively little R&D investment regarding feedstock, biopower conversion, and infrastructure	• Considered a third-generation bioenergy source • Mainly considered for biofuel purposes; however, some scientists are studying its biopower potential, both directly or via methane productiond
Agricultural Biomass and Animal Wastes					
Corn stover	7,587-7,967	Stover amounts could range from 3-4.5 dry tons/acre/year in fields producing 100-150 bushels of grain/acre	• Cultivation techniques are established • Using a resource that has previously gone unused • Stover conversion process could be added to grain-to-ethanol facilities	• Harvesting and transportation infrastructure not yet established • Excessive removal may lead to soil erosion and nutrient runoff • Requires high level of nutrients and fertile soils	• Corn grain and stover use has reinvigorated the food-fuel debate • Can be used for gasification, combustion, or pyrolysis technologies for electricity or biochemical processes for biofuels

Appendix A. (Continued).

Feedstock Type	Energy Value Btu/lb (dry)[a]	Feedstock Yield[b]	Selected Advantages	Selected Disadvantages	Comments[c]
Wheat straw	6,964-8,148	2.6 tons dry tons/acre	• Cultivation techniques are established • Using a resource that has previously gone unused	• Harvesting and transportation infrastructure not yet established • Excessive removal may lead to soil erosion and nutrient runoff	• Can be used for gasification, combustion or pyrolysis technologies to generate electricity or biochemical processes to biofuels
Sugarcane bagasse (residue once juice is extracted from the sugar cane; see above for sugarcane)	7,450-8,349	14%-30% of total sugarcane yield	• Sugarcane takes approximately one year to become established • Bagasse is collected as part of the main crop	• Bagasse availability limited to a few states in the South and Hawaii • Ash content can be high	• Literature mostly centers on its use for ethanol • The bagasse is used to power sugarcane mills in many parts of the world.
Cattle manure	8,500	Based on manure excretion rate of cow	• Using a resource that is generally regarded as a waste product with little to no value • Using a resource that has undesirable environmental impacts if improperly managed • Collection systems established for dairy manure	• Technology to convert manure to electricity is expensive • Difficult for some agricultural producers to sell power to utilities due to economics and utility company collaboration	• Well suited for anaerobic digestion to generate electricity

Feedstock Type	Energy Value Btu/lb (dry)[a]	Feedstock Yield[b]	Selected Advantages	Selected Disadvantages	Commentsc
			• Water and air quality improvement		
Industrial Biomass					
Municipal solid waste (MSW)	5,100 (on an as arrived basis)	1,643 lbs/person/year	• A resource available in abundant supply • Diverts MSW from landfill disposal • Well-commercialized technology (waste-to-energy plants)	• Could serve as a disincentive to separate and recycle certain waste • Air emissions are strictly regulated to control the release of toxic materials often in MSW; toxins removed from air emissions will be transferred to waste ash, which may require disposal as hazardous waste • Costs are substantially higher than landfill in most areas	• Not considered by some as a renewable energy feedstock because some of the waste materials are made using fossil fuels • Well suited for combustion (waste to energy plants), gasification, pyrolysis, or anaerobic digestion technologies to generate electricity
Fossil Fuels					
Coal (low rank; lignite/sub-bituminous)	6,437-8,154	Not applicable	• Established infrastructure • Reliable	• Limited resource • Major source of mercury, SO2, and NOx emissions	
			• Relatively inexpensive	• Main source of U.S. greenhouse gas emissions • Generates a tremendous amount of waste ash that likely contains a host of hazardous constituents	

Appendix A. (Continued).

Feedstock Type	Energy Value Btu/lb (dry)[a]	Feedstock Yield[b]	Selected Advantages	Selected Disadvantages	Comments[c]
Coal (high rank; bituminous)	11,587-12,875	Not applicable	• Established infrastructure • Reliable • Relatively inexpensive	• Limited resource • Major source of mercury, SO2 and NOx emissions • Main source of U.S. greenhouse gas emissions • Generates a tremendous amount of waste ash that likely contains a host of hazardous constituents	
Oil (typical distillate)	18,025-19,313	Not applicable	• Established infrastructure • Reliable	• Limited resource • Major source of SO2 and NOx emissions • Purchased in large quantities from foreign sources	

Source: Compiled from various sources by CRS and Lynn Wright, biomass consultant working with Oak Ridge National Laboratory.

Notes: The information provided in this table are estimates for general use. Multiple factors including location, economics, and technical parameters will influence the data on a case-by-case basis. Lynn Wright, biomass consultant working with Oak Ridge National Laboratory, provided the following comments: The infrastructure to handle woody resources (both forest residues and plantation grown wood) already exists in the pulp and paper industry and can be easily used for the bioenergy industry. Most woody biomass resources (whether forest residues or plantation grown wood) will be delivered as chips similar to current pulp and paper industry practices. However, new equipment and harvest techniques may allow delivery as bundles or whole trees in some situations. Wood resources such as chipped pine (softwoods) and hardwoods and urban wood residues are already being used to generate electricity using direct combustion technologies, all woody feedstocks are well suited for all thermal conversion technologies including combustion, gasification and pryolysis to generate electricity. Biopower can also be produced from the black liquor by-product of both pulp and ethanol production. Clean wood chips from willow, hybrid poplar, and other hardwoods are also very suitable for conversion to liquid fuels using biochemical conversion technologies.

[a] Energy values for the following feedstocks were obtained from Oak Ridge National Laboratory, Biomass Energy Data Book: Edition 2, ORNL/Tm-2009/098, December 2009, http://cta.ornl.gov/bedb/pdf/BEDB2_Full_Doc.pdf; Table A.2 "Heat Content Ranges for Various Biomass Fuels"; willow, hybrid poplar, pine = Forest Residues - softwoods, switchgrass, miscanthus (converted from kj/kg to Btu/lb) corn stover, sugarcane bagasse and wheat straw. Energy values for fossil fuels were obtained by converting the heating values (GJ/t) provided in Jonathan Scurlock, *Bioenergy Feedstock Characteristics*, Oak Ridge National Laboratory, 2002, http://bioenergy.ornl.gov/ papers/misc/biochar_factsheet.html to an energy value (Btu/lb). The energy value for sawmill residue was obtained from Nathan McClure, Georgia Forestry Commission, "Forest Biomass as a Feedstock for Energy Production," oral presentation for Georgia Bioenergy Conference, August 2, 2006, http://www.gabioenergy.org/ppt/McClure—Forest%20Biomass%20as%20a%20Feedstock%20for%20Energy%20Production.pdf. The energy value for algae was obtained from Oilgae, "Answers to some Algae Oil FAQs—Heating Value, Yield ...," February 2007, *http://www.oilgae.com/blog/2007/02/answers-to-some-algae-oil-*. The energy value of manure on a dry ash-free basis was obtained from Texas Cooperative Extension, *Manure to Energy: Understanding Processes, Principles and Jargon*, E-428, 2006, http://tammi.tamu.edu/ManurtoEnrgyE428.pdf. The manure heating value may be reduced by the ash and moisture content of the manure given certain conditions. The energy value of municipal solid waste was obtained from C. Valkenburg, C.W. Walton, and B.L. Thompson, et al., *Municipal Solid Waste (MSW) to Liquid Fuels Synthesis, Volume 1: Availability of Feedstock and Technology*, Pacific Northwest National Laboratory, PNNL-18144, December 2008, http://www.pnl.gov/main/publications/external/technical_reports/PNNL-18144.pdf. The energy value for sorghum was obtained using a value for sudan grass, a closely related crop, from the European PHYLLIS database http://www.ecn.nl/phyllis/dataTable.asp.

[b] The harvest frequency is on an annual basis unless stated otherwise. Energy yield ranges for willows, poplars, pines, switchgrass, miscanthus, sugarcane, sugarcane bagasse and sorghum were provided by Lynn Wright, biomass consultant working with Oak Ridge National Laboratory. Energy yields for miscanthus. and switchgrass were also discussed with Jeffrey Steiner (USDA), August 2010. Energy yields for hybrid poplar were also obtained from Minnesota Department of Agriculture, *Minnesota Energy from Biomass*, http://www.mda.state Energy yield for pine chips (forest residues) was obtained from calculations from data in David A. Hartman et al., *Conversion Factors for the Pacific Northwest Forest Industry* (Seattle, WA; Univ. of Washington, Institute of Forest Products, no date), pp. 6, 47. Energy yield for corn stover was obtained from R.L Nielsen, *Questions Relative to Harvesting & Storing Corn Stover*, Purdue University, AGRY-95-09, September 1995, http://www.agry.purdue.edu/ext/corn/pubs/agry9509.htm. Energy yield for wheat straw was obtained from Jim Morrison, Emerson Nafziger, and Lyle Paul, *Predicting Wheat Straw Yields in Northern Illinois*, University of Illinois at Urbana-

Champaign, 2007, http://cropsci.illinois.edu/research/rdc/dekalb/ publications/2007/PredictingWheatStrawYieldsFinalReportToExtensionMay2007.pdf; In general, it is assumed a dairy cow excretes 150lbs of manure/day based on the American Society of Agricultural and Biological Engineers (ASABE) Manure Production and Characteristics Standard D384.2, March 2005. Energy yield for municipal solid waste was calculated based on data from U.S. Environmental Protection Agency Office of Solid Waste http://www.epa.gov/osw/basic-solid.htm (In 2008, U.S. residents, businesses, and institutions produced about 250 million tons of MSW, which is approximately 4.5 pounds of waste per person per day). Energy yield for miscanthus in Europe was obtained from Clifton-Brown, J.C., Stampfl, P.A., and Jones, M.B., *Miscanthus Biomass Production for Energy in Europe and Its Potential Contribution to Decreasing Fossil Fuel Carbon Emissions*. Global Change Biology, 10, (2004) pp. 509-518; Energy yield for siwtchgrass was obtained from McLaughlin, S.B., and Kszos, L.A., "Development of Switchgrass (panicum virgatum) as a bioenergy feedstock in the United States." Biomass and Bioenergy 28 (2005) pp. 515-535. Energy yield for sorghum was obtained from W.L. Rooney, et al, "Designing Sorghum as a Dedicated Bioenergy Feedstock." *Biofuels, Bioproducts, and Biorefining*. 1, (2007) pp.147-157; Energy yield for sugarcane/energycane obtained from http://www.ars.usda.gov/research/publications/publications.htm?seq_no_115=251543&pf=1 (a web-published abstract of a book chapter written by Bransby et. al. and submitted for publication in February 2010); Energy yield for sugarcane baggase was obtained from http://www.ars.usda.gov/research/publications/publications.htm?seq_no_115=254594&pf=1 (an abstract of a book chapter prepared by R. Viator, P. White, and E. Richard, and entitled " Sustainable Production of Energycane for Bio-energy in the Southeastern U.S." submitted for publication by the Sugarcane Research Unit in Houma, LA in August 2010).

[c] For more information on the state of combustion, pyrolysis, gasification, and anaerobic digestion technologies, see the shaded text box on page 5.

[d] For more information, see Stanford University, "Stanford Researchers Find Electrical Current Stemming from Plants," press release, April 13, 2010, http://news.stanford.edu/news/2010/april/electric-current-plants and John Ferrell and Valerie Sarisky-Reed, *National Algal Biofuels Technology Roadmap*, U.S. D epartment of Energy Office of Energy Efficiency and Renewable Energy Office of the Biomass Program, May 2010, http://www1.eere.energy algal_biofuels_roadmap.pdf.

Questions remain about what would be needed to increase biopower production and simultaneously address technological, environmental, and agricultural concerns. Because market and regulatory uncertainties exist for biopower, the agricultural community may be hesitant to grow the amount of biomass feedstocks needed to support large-scale biopower production. Moreover, most biopower technologies, with the exception of combustion and co-firing systems, have yet to reach commercial status. Improvements to the remaining biopower conversion technologies may arise when there is a solid market for biopower. Furthermore, regulatory uncertainty has contributed to the reluctance to develop biopower. There is no federal mandate requiring the production of biopower, although more than 25 states have implemented state renewable portfolio standards or goals that include biopower. Additional assurances of federal support, whether technical, economic, or through renewable mandates, could spur commitments by investors, the technology community, and others.

Appendix B. Biopower R&D Authorizations

R&D Authorizations

Congress has enacted numerous provisions that authorize the Departments of Energy (DOE) and Agriculture (USDA) to conduct biopower research, development, and demonstration projects (RD&D) and to support biopower commercial application efforts.[26] At least eight public laws contain one or more biopower provisions:

- P.L. 95-620, Powerplants and Industrial Fuel Use Act of 1978
- P.L. 96-294, Energy Security Act of 1980
- P.L. 106-224, Biomass Research and Development Act of 2000
- P.L. 107-171, Farm Security and Rural Investment Act of 2002
- P.L. 108-148, Healthy Forest Restoration Act of 2003
- P.L. 109-58, Energy Policy Act of 2005
- P.L. 110-140, Energy Independence and Security Act of 2007
- P.L. 110-246, Food, Conservation, and Energy Act of 2008

The public laws discussed in this section are summaries of provisions at the time of enactment to illustrate the evolution of bioenergy policy. Some provisions may have been amended since enactment. A comprehensive legislative history of current law is beyond the scope of this report.

1978-1980: Biopower Legislative Origin

Both the Powerplant and Industrial Fuel Use Act of 1978 (P.L. 95-620) and the Energy Security Act of 1980 (P.L. 96-294) introduced the concept of biopower to the legislative arena. However, the enacted legislation emphasized the use of biomass as a liquid fuel to reduce dependence on imported petroleum and natural gas. Biomass used to generate electricity appears to have received less legislative support compared to biomass use as a liquid fuel, based on the report language and authorizations.

Powerplant and Industrial Fuel Use Act of 1978 (P.L. 95-620)

The legislative origin of biopower stems from the Powerplant and Industrial Fuel Use Act of 1978. The act aimed to restrict the use of oil and natural gas as fuel in an attempt to mitigate the oil crisis of the mid-1970s by encouraging industries and utilities to reduce oil use. It required new power plants to operate using coal or alternate fuel sources. Otherwise, the act did not provide explicit support for biopower RD&D and commercial application.

- § 103(a)(6) - defines alternate fuel, in part, as electricity or any fuel, other than natural gas or petroleum, from sources such as biomass, municipal, industrial or agricultural wastes, wood, and renewable and geothermal energy sources.

Energy Security Act of 1980 (P.L. 96-294)

- § 203(4)(B) - defines biomass energy, in part, as energy or steam derived from the direct combustion of biomass for the generation of electricity, mechanical power, or industrial process heat.
- § 203(5)(B) - defines biomass energy project, in part, as any facility (or portion of a facility) located in the United States which is primarily for the combustion of biomass for generating industrial process heat, mechanical power, or electricity, including cogeneration.
- § 203(19) - defines a small-scale biomass energy project as a biomass energy project with an anticipated annual production capacity of not more than 1 million gallons of ethanol per year, or its energy equivalent of other forms of biomass energy.

- § 211(a) - requires DOE and USDA to collaborate on a biomass energy production and use plan and on providing financial assistance for biomass energy projects.
- § 251(a) - indirect reference to biopower; stipulates the establishment of demonstration biomass energy facilities by the Secretary of Agriculture to exhibit the most advanced technology available for producing biomass energy.
- § 252 - indirect reference to biopower; modifies § 1419 of the National Agricultural Research, Extension, and Teaching Policy Act of 1977 (P.L. 95-113) to better address biomass energy for RD&D purposes; authorizes the Secretary of Agriculture to award grants for research related to, in part, the development of the most economical and commercially feasible means of producing, collecting, and transporting agricultural crops, wastes, residues, and byproducts for use as feedstocks for the production of alcohol and other forms of biomass energy.
- § 255(a) - indirect reference to biopower; adds a Biomass Energy Educational and Technical Assistance Program to Subtitle B of P.L. 95-113 to provide technical assistance to producers for efficient use of biomass energy and disseminate research results to producers about biomass energy, among other things.

1981-1999: Biopower Legislation and Technology

Congress did not significantly address biopower during most of the 1980s and 1990s partially due to stable conventional energy prices and supplies. Some biopower technologies emerged during this time period with low success rates due to poor design and inadequate management (e.g., anaerobic digestion systems). Other reliable biopower technologies were developed during this time period (e.g., biomass co-firing), but these could not compete economically with other energy sources.

2000-Present: Biopower Legislative Action

Described below are a variety of biopower provisions contained in public laws since 2000. Although many of the provisions focus primarily on the use of biomass for liquid transportation fuel, there is also legislative support for biopower. Both DOE and USDA have the authority to conduct RD&D and support commercial application efforts for biopower. However, project summaries and financial allotments indicate the majority of resources in recent years were directed toward liquid fuels for transportation.[27]

Biomass Research and Development Act of 2000 (P.L. 106-224)

The Biomass Research and Development Act[28] established a partnership between the USDA and DOE for RD&D on the production of biobased industrial products. (This act was amended by the Energy Policy Act of 2005, P.L. 109-58.) The original provisions included:

- § 303(2) - defines biobased industrial products to include fuels, chemicals, building materials, or electric power or heat produced from biomass.
- § 305 - implicit reference to biopower; establishes the Biomass Research and Development Board to coordinate research and development activities relating to biobased industrial products; Board membership includes a representative from DOE, USDA, Department of the Interior, the U.S. Environmental Protection Agency, the National Science Foundation, and the Office of Science and Technology Policy.
- § 306 - implicit reference to biopower; establishes the Biomass Research and Development Technical Advisory Committee to, in part, advise the Biomass Research and Development Board concerning the technical focus and direction of requests for proposals issued under the Biomass Research and Development Initiative
- § 307 - implicit reference to biopower; authorizes the Secretaries of Agriculture and of Energy to, in part, competitively award grants, contracts, and financial assistance to eligible entities that can perform research on biobased industrial products. For example, grants may be rendered to an entity conducting research on advanced biomass gasification and combustion to produce electricity (§ 307(d)(2)(e)); related research in advanced turbine and stationary fuel cell technology for production of electricity from biomass (§ 307(d)(2)(f)); biomass gasification and combustion to produce electricity (§ 307(d)(3)(A)(v)); and any research and development in technologies or processes determined by the Secretaries, acting through their respective points of contact and in consultation with the Biomass Research and Development Board (§ 307(d)(4)).

Farm Security and Rural Investment Act of 2002 (P.L. 107-171)

- § 9003 - authorizes the Secretary of Agriculture to award grants to assist in paying the development and construction costs of biorefineries in order to carry out projects that demonstrate their commercial viability for converting biomass to fuels or chemicals.

- § 9003(b)(2) - defines biorefinery as equipment and processes that convert biomass into fuels and chemicals; and may produce electricity.

Healthy Forest Restoration Act of 2003 (P.L. 108-148)

- § 203 - establishes the Biomass Commercial Utilization Grant Program; authorizes the Secretary of Agriculture to make grants to the owner or operator of a facility that uses biomass as a raw material to produce one or more of several outputs, including electric energy.

Energy Policy Act of 2005 (EPAct05; P.L. 109-58)

- § 931(f) - authorizes the Secretary of Energy, in consultation with the Secretary of Agriculture, to implement rural demonstration projects that use renewable energy technologies to assist in delivering electricity to rural and remote locations from biomass.
- § 932 (b)(1) - authorizes the Secretary of Energy to conduct a program of RD&D, and commercial application for bioenergy including biopower energy systems.
- § 932 (d)(B)(iv) - authorizes the Secretary of Energy to demonstrate the commercial application of integrated biorefineries from the commercial application of biomass technologies for energy in the form of electricity or useful heat.
- § 941(a) - amends the definition for biobased product in P.L. 106-224 to mean an industrial product (including chemicals, materials, and polymers) produced from biomass, or a commercial or industrial product (including animal feed and electric power) derived in connection with the conversion of biomass to fuel.
- § 941(d)(1) - modifies membership of the Biomass Research and Development Technical Advisory Committee (P.L. 106-224 § 306); replaces an individual affiliated with the biobased industrial products industry with an individual affiliated with the biofuels industry; adds an individual affiliated with the biobased industrial and commercial products industry; requires committee members as described in P.L. 106-224, § 306(b)(1)(C), (D), (G), and (I) to have expertise in 'fuels and biobased products' whereas previously members were to have expertise in 'biobased industrial products'.

- § 941(e)(1) - modifies the Biomass Research and Development Initiative (P.L. 106-224, § 307(a)) to focus on "research on, and development and demonstration of, biobased fuels and biobased products, and the methods, practices and technologies, for their production." Previously the initiative focus was on "research on biobased industrial products."
- § 941(e)(2) - adds to the Biomass Research and Development Initiative (P.L. 106- 224, § 307) an objectives section and a technical areas section, in addition to other sections, that specify biobased fuels as a priority. For example, the initiative is to support "product diversification through technologies relevant to production of a range of biobased products (including chemicals, animal feeds, and cogenerated power) that eventually can increase the feasibility of fuel production in a biorefinery."

Energy Independence and Security Act of 2007 (EISA; P.L. 110-140)

- § 231(1) - modifies EPAct05 § 931(b) by adding an authorization of $963 million for FY2010. Section 931 of the EPAct05 authorizes the Secretary of Energy to conduct programs of renewable energy RD&D, and commercial application.
- § 231(2) - modifies EPAct05 § 931(c)(2) to increase authorized funding for FY2008 from $251 million to $377 million; Also modifies EPAct05 § 931(c)(3) to increase authorized funding for FY2009 from $274 million to $398 million.

The Food, Conservation, and Energy Act of 2008

(2008 Farm Bill, P.L. 110-246)

- § 7526 - reauthorizes the Sun Grant program, which requires USDA to coordinate with DOE and land-grant colleges and universities to provide grants to the Sun Grant centers to enhance the efficiency of bioenergy and biomass research and development programs.
- § 9001 - defines biorefinery as a facility that converts renewable biomass into biofuels and biobased products; and may produce electricity.
- § 9008 - defines biobased product as an industrial product (including chemicals, materials, and polymers) produced from biomass, or a

commercial or industrial product (including animal feed and electric power) derived in connection with the conversion of biomass to fuel.

- § 9011 - establishes the Biomass Crop Assistance Program which provides financial assistance to producers or entities that deliver eligible biomass material to designated biomass conversion facilities for use as heat, power, biobased products or biofuels.
- § 9012 - authorizes the Secretary of Agriculture, acting through the Forest Service, to conduct a competitive R&D program to encourage use of forest biomass for energy.
- § 9013(a)(2) - defines a community wood energy system as an energy system that primarily services public facilities owned or operated by state or local governments, including schools, town halls, libraries, and other public buildings; and uses woody biomass as the primary fuel. The term includes single facility central heating, district heating, combined heat and energy systems, and other related biomass energy systems.
- § 9013(b) - establishes the Community Wood Energy Program and authorizes the Secretary of Agriculture, acting through the Forest Service, to provide grants of up to $50,000 for up to 50% of the cost for communities to plan and install wood energy systems in public buildings.

End Notes

[1] The Renewable Fuel Standard, a mandate to ensure that domestic transportation fuel contains a specified volume of biofuels, is one reason most legislative and administrative efforts have focused on development of biofuels for transportation. For more information, see CRS Report R40155, *Renewable Fuel Standard (RFS): Overview and Issues*, by Randy Schnepf and Brent D. Yacobucci.

[2] U.S. Energy Information Administration, *Annual Energy Review 2009*, DOE/EIA-0384(2009), August 2010, http://www.eia.doe.gov/aer/elect.html.

[3] U.S. Energy Information Administration, *Renewable Energy Annual 2008 Edition*, August 2010, http://www.eia.doe.gov/cneaf/solar.renewables/page/rea_data/rea_sum.html. Biopower constituted roughly 14.4% of electricity generation from renewable energy sources in 2008, preceded by conventional hydroelectric power and wind, which constituted roughly 67% and 14.5%, respectively.

[4] U.S. Energy Information Administration, *Annual Energy Outlook 2010*, DOE/EIA-0383(2010), Washington, DC, April 2010, http://www.eia.doe.gov/oiaf/aeo/. The bulk of this increase is expected to come from growth in co-firing operations. Co-firing is the combustion of a supplementary fuel (e.g., biomass) and coal concurrently.

[5] Pew Center on Global Climate Change , *Biopower*, December 2009,

http://www.pewclimate.org/docUploads/ Biopower%20final%2011%2009.pdf. Certain analysis indicates that feedstock supply should be located within a 50- mile radius to avoid excessive transportation costs: Marie E. Walsh, Robert L. Perlack, and Anthony Turhollow et al., *Biomass Feedstock Availability in the United States: 1999 State Level Analysis*, Oak Ridge National Laboratory, January 2000, http://bioenergy.

[6] Executive Order 13514 defines sustainability as the creation and maintenance of conditions that allow humans and animals to exist in productive harmony, and that permit fulfilling the social, economic, and other requirements of present and future generations. For more information, see CRS Report R40974, *Executive Order 13514: Sustainability and Greenhouse Gas Emissions Reduction* , by Richard J. Campbell and Anthony Andrews.

[7] For more information on biomass definitions, see CRS Report R40529, *Biomass: Comparison of Definitions in Legislation Through the 111th Congress*, by Kelsi Bracmort and Ross W. Gorte.

[8] Some of this information may be provided in a forthcoming update to the frequently cited DOE/USDA Billion-Ton Study, *Biomass as Feedstock for a Bioenergy and Bioproducts Industry: The Technical Feasibility of a Billion-Ton Annual Supply*, April 2005, http://www1.eere.energy.

[9] In September 2010 the National Renewable Energy Laboratory released a comprehensive mapping application that may provide better data to compare biomass feedstock and biopower by location. National Renewable Energy Laboratory, "NREL Releases BioEnergy Atlas—A Comprehensive Biomass Mapping Application," press release, September 28, 2010, http://www.nrel.gov/news/press/2010/891.html.

[10] CRS calculations based on 2008 total U.S. retail electricity sales available at http://www.eia.doe.gov/electricity/esr/ esr_sum.html. Power plant capacity factor was assumed to be 80% with 988 growing acres required per megawatt; see http://bioenergy. The yield, six dry tons/acre, is similar to what may be achieved by switchgrass. Land in farms data for Iowa obtained from the 2007 Census of Agriculture, available at *.http://www. agcensus.usda.gov/Publications/2007/Online_Highlights/County_Profiles/Iowa*

[11] Peter McKendry, "Energy Production from Biomass (Part 1): Overview of Biomass," *Bioresource Technology*, vol. 83 (2002), pp. 37-46.

[12] Peter McKendry, "Energy Production from Biomass (Part 1): Overview of Biomass," *Bioresource Technology*, vol. 83 (2002), pp. 37-46.

[13] Section 201 of the Energy Independence and Security Act of 2007 (EISA; P.L. 110-140) defines lifecycle emissions as follows: "(H) LIFECYCLE GREENHOUSE GAS EMISSIONS.—The term 'lifecycle greenhouse gas emissions' means the aggregate quantity of greenhouse gas emissions (including direct emissions and significant indirect emissions such as significant emissions from land use changes), as determined by the Administrator, related to the full fuel lifecycle, including all stages of fuel and feedstock production and distribution, from feedstock generation or extraction through the distribution and delivery and use of the finished fuel to the ultimate consumer, where the mass values for all greenhouse gases are adjusted to account for their relative global warming potential." 42 U.S.C. §7545(o)(1). For more information on lifecycle emissions, see CRS Report R40460, *Calculation of Lifecycle Greenhouse Gas Emissions for the Renewable Fuel Standard (RFS)*, by Brent D. Yacobucci and Kelsi Bracmort.

[14] For more information on carbon neutrality of biomass energy, see CRS Report R41603, *Is Biopower Carbon Neutral?*, by Kelsi Bracmort.

[15] The rule sets thresholds for GHG emissions that define when permits are required for new and existing industrial facilities. For more information on the history of the Tailoring Rule, see CRS Report R41103, *Federal Agency Actions Following the Supreme Court's Climate Change Decision: A Chronology*, by Robert Meltz..

[16] EPA's decision on biomass combustion and biogenic activities is described in further detail on pages 419-422 of the final rule, available at http://www.epa.gov/nsr/documents/ 20100413final.pdf. For more information on the final rule, see CRS Report R41212, *EPA Regulation of Greenhouse Gases: Congressional Responses and Options*, by James E. McCarthy and Larry Parker.

[17] Environmental Protection Agency, "EPA to Defer GHG Permitting Requirements for Industries that Use Biomass/Three-year deferral allows for further examination of scientific and technical issues associated with counting these emissions ," press release, January 12, 2011, http://yosemite.epa.gov/opa/admpress.nsf/ d0cf6618525a9efb85257359003fb69 d/4369c709163915b485257816005971bb!OpenDocument. Biogenic includes facilities that emit CO2 from sources originating via a biological processes, such as landfills.

[18] BACT is an emissions limitation that is based on the maximum degree of control that can be achieved. It is a caseby-case decision that considers energy, environmental, and economic impact. BACT can be add-on control equipment or modification of the production processes or methods. BACT may be a design, equipment, work practice, or operational standard if imposition of an emissions standard is infeasible. Environmental Protection Agency, PSD and Title V Permitting Guidance For Greenhouse Gases, November 2010, *http://www. epa.gov/nsr/ghgdocs/epa-hq-oar2010-0841-0001.pdf.*

[19] Environmental Protection Agency, *PSD and Title V Permitting Guidance For Greenhouse Gases*, November 2010, http://www.epa.gov/nsr/ghgdocs/epa-hq-oar-2010-0841-0001.pdf.

[20] 26 U.S.C. § 45.

[21] 26 U.S.C. § 48.

[22] U.S. Department of the Treasury, *Payments for Specified Energy Property in Lieu of Tax Credits under the American Recovery and Reinvestment Act of 2009*, January 2011, http://www.treasury.gov/initiatives/recoveryguidance.pdf.

[23] For more information on BCAP, see CRS Report R41296, *Biomass Crop Assistance Program (BCAP): Status and Issues*, by Megan Stubbs. BCAP provides financial assistance to producers or entities that deliver eligible biomass material to designated biomass conversion facilities for use as heat, power, biobased products, or biofuels.

[24] For more information on the proposed RES in S. 1462, see CRS Report R40837, *Summary and Analysis of S. 1462: American Clean Energy Leadership Act of 2009, As Reported*, coordinated by Mark Holt and Gene Whitney.

[25] For more information, see CRS Report R40890, *Summary and Analysis of S. 1733 and Comparison with H.R. 2454: Electric Power and Natural Gas*, by Stan Mark Kaplan.

[26] National Renewable Energy Laboratory, *Power Technologies Energy Data Book*, NREL/TP-620-39728, August 2006, *http://www.nrel.gov/analysis/power_databook/docs/pdf/39728_ complete.pdf.*

[27] For information on biomass energy incentives, see CRS Report R40913, *Renewable Energy and Energy Efficiency Incentives: A Summary of Federal Programs*, by Richard J. Campbell, Lynn J. Cunningham, and Beth A. Roberts [28] The Biomass Research and Development Act is Title III of the Agricultural Risk Protection Act of 2000 (P.L. 106-224).

In: Carbon Considerations
Editors: W. Green and L. Wayman ISBN 978-1-62618-435-0

Chapter 6

IS BIOPOWER CARBON NEUTRAL?*

Kelsi Bracmort

SUMMARY

Congress has expressed interest in biopower—electricity generated from biomass. Biopower, a baseload power source, has the potential to strengthen rural economies, enhance energy security, and improve the environment, proponents say. Biopower could be produced from a large range of biomass feedstocks nationwide (e.g., urban, agricultural, and forestry wastes and residues). One challenge to biopower production is a readily available feedstock supply. At present, biopower requires tax incentives to be competitive with conventional fossil fuels. If Congress considers a renewable electricity standard or other measures (e.g., farm bill energy programs) that include biopower, there may be concerns about the carbon neutrality of biopower. Congressional support for biopower has aimed to promote energy diversity and improve energy security, and has generally assumed that biopower is carbon neutral. An energy production activity is typically classified as carbon neutral if it produces no net increase in greenhouse gas (GHG) emissions on a life-cycle basis. The premise that biopower is carbon neutral has come under scrutiny as its potential to help meet U.S. energy demands and reduce U.S. greenhouse gas emissions is more closely examined.

* This is an edited, reformatted and augmented version of a Congressional Research Service publication, CRS Report for Congress R41603, from www.crs.gov, prepared for Members and Committees of Congress, dated January 2, 2013.

Whether biopower is carbon neutral depends on many factors, including the definition of carbon neutrality, the feedstock type, the technology used, and the time frame examined. Carbon flux (emission and sequestration) varies at each phase of the biopower pathway, given site- and operation-specific factors. A life-cycle assessment (LCA) is a common technique to calculate the environmental footprint, including the carbon flux, of a particular biopower pathway. However, past legislation has not required a standardized LCA.

Interest in the carbon classification of biopower is in part due to sustainability and air quality concerns. Where the feedstock supply for biopower originates, if it is managed in a sustainable manner, and whether the associated air quality impacts from biopower generation are tolerable are questions that are part of the biopower carbon-neutrality debate. Congress may decide whether the current carbon-neutral designation for biopower is accurate, or whether additional carbon accounting for biopower is warranted and what impact this accounting might have on renewable energy, agricultural, and environmental legislative goals.

Rulings by the U.S. Environmental Protection Agency have raised questions about the carbon neutrality of biopower. For instance, the 2010 Prevention of Significant Deterioration and Title V Greenhouse Gas Tailoring Rule did not exempt emissions from biomass combustion. Some view EPA's decision as equating biomass emissions with fossil fuel emissions. EPA decided in 2011 to defer for three years GHG permitting requirements for carbon dioxide emissions from bioenergy and other biogenic stationary sources in order to conduct a detailed examination of the science associated with these emissions. EPA's Scientific Advisory Board conducted an independent review of the agency's biogenic accounting framework and released its findings in September 2012. The board acknowledged the "daunting task" of assessing the greenhouse gas implications of bioenergy, and the "narrow regulatory boundaries" within EPA's purview that limit the consideration of greenhouse gas flux at various points along the bioenergy pathway.

State perspectives on the tailoring rule are divided. Some states contend that treating biomass combustion the same as fossil fuel combustion will result in excessive permitting requirements and fees that jeopardize renewable energy development. Other states argue that not treating it the same will aggravate climate change over time.

INTRODUCTION

Biomass energy, or bioenergy, is receiving increased attention as an alternative to fossil fuel energy because of its potential to improve the environment, provide energy security, and promote economic development. Biomass is organic matter—woody biomass, agricultural biomass, animal wastes, aquatic biomass—that can be converted to energy (e.g., heat, electricity, or liquid transportation fuels). A substantial supply of biomass feedstocks may be required to produce bioenergy, depending on current and future renewable energy mandates. A large biomass feedstock requirement could lead to water, land use, sustainability, and economic concerns.[1] Federal support for bioenergy is available via tax incentives, loan guarantees, technical and financial assistance, and mandated use requirements.

Legislative support for bioenergy, particularly for biopower (electricity generated from biomass),[2] has thus far been granted under the premise that it is carbon neutral. As more public and private resources are spent—or as greater spending is debated—on biopower, more attention is being directed to the rationale for designating biopower as carbon neutral. The carbon-neutral designation is typically assigned to an energy production activity that essentially produces no net increase in greenhouse gas (GHG) emissions on a life-cycle basis (or one in which the amount of carbon dioxide emitted during the power production cycle is absorbed).[3] Where biopower stands among the other renewable energy sources with respect to GHG emissions may affect the level of future legislative support granted to it. Is biopower carbon neutral? If it is not, should it receive the same type and amount of federal resources as carbon-neutral or less carbon-intensive energy sources? How might the federal government account for carbon associated with biopower ventures? How Congress addresses biopower's carbon neutrality could shape how Congress treats biopower in general.

Many views exist about whether biopower is carbon neutral. Some contend that biopower is carbon neutral because the carbon released during bioenergy production comes from a carbon-neutral feedstock—biomass. Some argue that biopower is not carbon neutral because the amount of GHG emissions released per unit of energy during simple biopower combustion may be higher for certain biomass fuels than for fossil fuels; or, if the GHG emissions from certain biomass fuels are lower than fossil fuels, they are still not zero. These perspectives are often based on differing assumptions, technologies, and time frames.

The debate concerning biopower's designation as carbon neutral may intensify, given possible congressional and Administration decisions. Congress may consider legislation involving biopower (e.g., a renewable electricity standard, a clean energy standard, or the next farm bill). The U.S. Environmental Protection Agency (EPA) has already promulgated regulations involving biopower (e.g., the 2010 Prevention of Significant Deterioration and Title V Greenhouse Gas Tailoring Rule) and may consider further regulatory actions. Although EPA deferred for three years GHG permitting requirements for carbon dioxide emissions from biomass-fired and other biogenic sources,[4] Congress may still decide to examine how the widely accepted premise of biopower's carbon neutrality contributes to energy, environmental, and economic development goals.

This report discusses some of the most relevant factors to take into account when considering if biopower is carbon neutral. It does not discuss carbon accounting for other bioenergy pathways.[5]

BIOMASS CARBON CYCLE

The carbon cycle encompasses the many pathways through which carbon is exchanged between the atmosphere and the land and water.[6] Human activities (also called anthropogenic activities) contribute to the carbon cycle by emitting carbon dioxide (CO_2). The human contribution of CO_2 to the carbon cycle is relatively small compared to other contributions, but CO_2 released to the atmosphere from human activities is taken up by soils, vegetation, and the ocean at a rate that is relatively slower than the rate at which human activities are emitting CO_2. If the excess carbon is not stored in land and ocean sinks, the atmospheric concentration of CO_2 increases, potentially impacting the Earth's climate.

One significant anthropogenic source of CO_2 in the carbon cycle is energy production. The net effect of an energy activity on the carbon cycle can be classified in one of three ways. A "carbon-positive" activity releases CO_2 into the atmosphere. A "carbon-negative" activity removes more CO_2 from the atmosphere than it emits. A "carbon-neutral" activity is one where the CO_2 releases and absorption are in balance. There is no commonly accepted definition for a "carbon-neutral" activity in the biopower arena. Indeed, multiple assertions about carbon neutrality have been put forth by those involved with biomass energy, including the following:[7]

- Biomass energy is carbon neutral because biomass is naturally carbon neutral.
- Biomass energy is neutral if the activity removes as much CO_2 as was emitted into the atmosphere.

- Biomass energy is neutral only if the net life-cycle emissions are zero.[8]
- Biomass energy is neutral if it achieves lower net increases in atmospheric GHGs when compared to alternative energy activities.

There are pros and cons for each assertion. For instance, declaring that biomass energy is carbon neutral because biomass is naturally carbon neutral does not account for GHG emissions released due to management of crops grown for energy production (e.g., fertilizer). There may need to be additional plantings of certain biomass feedstocks to remove the CO_2 emitted from biomass cultivated for energy production.

The carbon cycles for a bioenergy system and a fossil fuel system differ in at least two ways: the carbon source (finite versus renewable) and the atmospheric carbon concentration (potentially stable versus additional; see *Figure 1*). Three main factors contribute to the amount of carbon emitted from biopower generation: feedstock production (cultivation and harvest), feedstock transport, and the biopower technology type. However, as noted by many sources, feedstock production also absorbs carbon during growth.

GREENHOUSE GAS ACCOUNTING FOR BIOPOWER PRODUCTION

Whether and how to conduct GHG accounting for biopower is an issue at the forefront of the bioenergy movement. GHG accounting can be used to compare the environmental footprint of a biopower operation with that of conventional fossil fuel (e.g, electricity from coal). The environmental footprint is often calculated using a life-cycle assessment (LCA), an analytic method for identifying, evaluating, and comparing the environmental impacts of emissions and the resource depletion associated with a specific process.[9] An LCA generally uses observed data and assumptions to model what GHGs are being released at each phase of the process. Ideally, an LCA would encompass economic and social factors for a more comprehensive assessment. Alternatively, an LCA can be one element used in assessing a preferred energy approach, along with cost and performance data. In some cases, even if LCA results favor a particular approach, an LCA alone might not be the deciding factor when choosing an energy process; financial objectives, policy goals, and other factors may influence which approach is selected.

An LCA has four major stages: goal and scope definition, inventory analysis, impact assessment, and interpretation.[10] While an LCA can be performed to meet various goals, the goals most applicable to the biopower arena are to support broad environmental assessments, support public policy, and provide information and direction to decision-makers. The goal and scope definition stage determines the amount of resources and time needed to conduct an LCA. The inventory analysis stage of an LCA quantifies energy and raw material requirements, atmospheric emissions, waterborne emissions, solid wastes, and other releases for the entire life cycle of a product, process, or activity. The impact assessment stage evaluates the potential human health and environmental impacts of the environmental resources and releases identified during the inventory analysis stage. The interpretation stage identifies, quantifies, checks, and evaluates the results of the inventory analysis and impact assessment stages, and communicates the results.

GHG accounting with an LCA can be performed at each phase of the biopower pathway: biomass cultivation and harvest, biomass transport, electricity generation, electricity transmission and distribution, and electricity end use (*Figure 2*). The first three phases of the biopower pathway (cultivation and harvest, transport, and electricity generation) are where the bulk of GHG emissions occur. GHG flux during the first three phases is site- and operation-specific, and depends on many factors, including the biomass type, management strategies, and the biopower generation technology.

Published LCAs for biopower are limited and, as noted above, may not be applicable to specific cases.[11] The LCAs performed are often tailored to one feedstock and one biopower technology type, and LCA results vary depending on assumptions such as the time frame of the assessment. The LCAs presented in this section are based on both a long time frame (e.g., cradle to grave) and a relatively short time frame (e.g., cradle to gate).[12] Different LCA time frames can lead to radically different, even contradictory, results.

Although biopower LCAs are scarce compared to liquid transportation biofuel LCAs, certain trends appear in existing assessments. For instance, the National Renewable Energy Laboratory (NREL) conducted a review and analysis of 57 biopower LCAs. The NREL review shows that biopower reduces greenhouse gas emissions when compared with fossil-based generation of electricity.[13]

And based on a separate review of more than 25 LCAs, biopower is in the top tier of bioenergy pathways that avoid the most GHG emissions and replace the largest amounts of fossil energy.[14] Approximately 15 of the LCAs reviewed included electricity as an end product, of which at least 10 had an LCA time frame of cradle to gate.

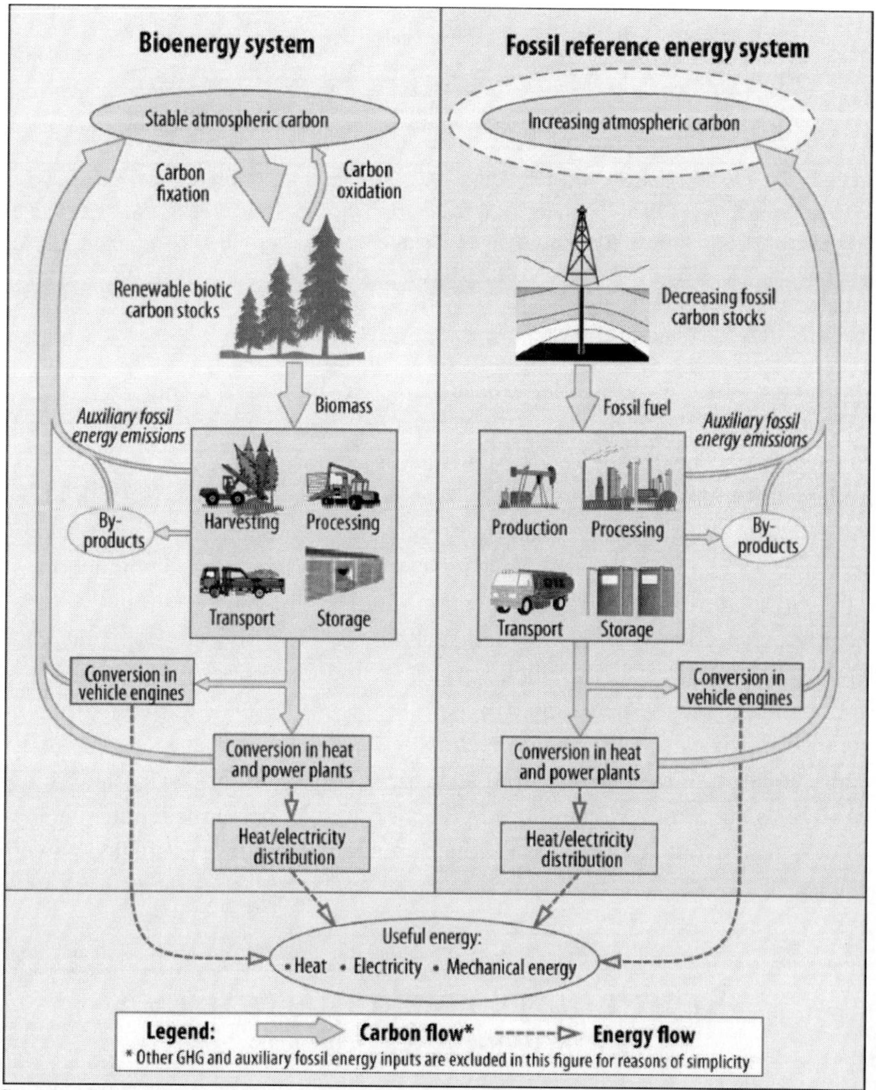

Source: International Energy Agency, IEA Bioenergy Task 38, Greenhouse Gas Balances of Bioenergy and Bioenergy Systems, 2002. Adapted by CRS.

Notes: The magnitude of the carbon flows, as indicated by the width of the arrows, is a significant part of the debate over the carbon neutrality of bioenergy.

Figure 1. Bioenergy CO2 Balance vs. Fossil Fuel CO2 Balance.

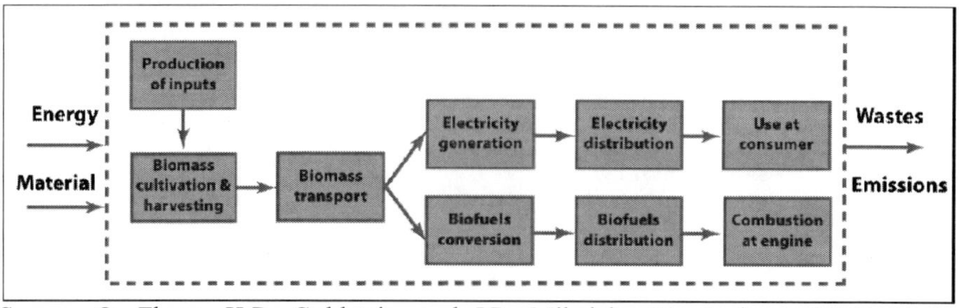

Source: Q. Zhang, K.R. Goldstein, and J.R. Mihelcic, "A Review of Life Cycle Assessment Renewable Energy Derived from Forest Resources," in Renewable Energy from Forest Resources in the United States, ed. B.D. Solomon, C. A. Luzadis (New York: Routledge, 2009). Adapted by CRS.

Figure 2. Biopower and Biofuel Pathways.

Biopower LCAs are usually limited to two biopower technology types: combustion and gasification. Both technologies have strengths and weaknesses.[15] The technology to co-fire (or combust) biomass with coal is available at commercial scale and is in use today. A cradle-to-grave DOE LCA found that a co-firing power plant, at both the 15% and 5% biomass levels (with a 350 and 354 MW plant capacity, respectively), compared with a power plant with no co-firing, led to a reduction in GHG emissions by roughly 18% and 5%, respectively.[16] Gasification technology is in the development and demonstration phase.[17] A cradle-to-grave DOE LCA for a hypothetical 113 MW biomass plant—performed in part to identify CO_2 emissions from a biopower plant using gasification technology—concluded that more than half of all net CO_2 emissions came from feedstock production, primarily the use of fossil fuels in farming operations.[18] The LCA was performed only on the biomass system and not for an immediate comparison with fossil-fueled power options.

Recent Developments Affecting Biopower Assessment

Certain actions have propelled the biomass carbon-neutrality issue to the center of the bioenergy debate. Most notable are the EPA's Prevention of Significant Deterioration (PSD) and Title V Greenhouse Gas Tailoring Rule, finalized in June 2010, and the June 2010 release of the Manomet Center for Conservation Sciences Biomass Sustainability and Carbon Policy Study.[19]

Also, many scientists contributed to the discussion with the submission of two letters to Members of Congress. The first letter provided the authors' reasons for supporting legislation that differentiates GHG emissions from bioenergy based on the source of the biomass.[20] The second letter provided the authors' reasons why equating biogenic carbon emissions (or bioenergy emissions) with fossil fuel emissions in the tailoring rule could hinder development of bioenergy technologies and may not be scientifically valid for certain sources of biomass.[21]

Title V Greenhouse Gas Tailoring Rule

The designation of biomass combustion as carbon neutral has come under scrutiny partly due to the tailoring rule finalized by EPA. The tailoring rule was developed by EPA to limit the number of facilities that would be required to obtain permits for greenhouse gas emissions under the Clean Air Act's Prevention of Significant Deterioration (PSD) and Title V permit programs.[22] By statute, the PSD program requires any new or modified facility emitting more than 100 tons or 250 tons of pollutants subject to regulation to obtain a permit prior to construction or operation. Title V sets a statutory 100-ton threshold for operating permits and applies to existing as well as new facilities. These thresholds limit the applicability of the permit requirements to the largest emitters of most pollutants that have traditionally been covered by the Clean Air Act; but when applied to carbon dioxide emissions, they would not serve the same limiting function. EPA estimates that a 100-ton threshold for GHGs would ultimately require more than 6 million sources to obtain Title V permits, with administrative costs for the state and local permitting authorities in the billions of dollars annually. Faced with these "absurd results" of following the letter of the law, EPA developed the tailoring rule to focus first on the largest emitters of GHGs. The rule sets a 75,000 ton per year of carbon dioxide equivalent (CO_2e)[23] threshold initially, allowing the agency to decide over a six-year period what to do about smaller sources.[24]

The tailoring rule does not exempt emissions from biomass combustion.[25] The rule grants exemptions based not on source category (e.g., fossil fuels, biomass) but on carbon tonnage emitted from a facility. One reason EPA did not exempt the biomass industry from the tailoring rule requirements is lack of information demonstrating the costs and administrative burdens the biopower industry would face if subject to the permitting requirements.[26] EPA issued a call for information in July 2010 to request comment on possible accounting approaches for biogenic emissions under the tailoring rule.[27] EPA received over 7,000 comments in response (approximately 600 were detailed responses). The two central themes for the comments reviewed by EPA included the request for certainty when it comes to biomass energy, and the recognition that all biomass is not the same.[28]

In July 2011, EPA announced that it would defer for a period of three years GHG permitting requirements for CO_2 emissions from biomass-fired and other biogenic sources.[29] Some are concerned the deferment may lead to delays in biomass energy investments, due to uncertainty about how EPA will address these types of facilities in three years. However, data is not available to determine if the ruling is inhibiting investment. EPA is using the time to conduct a detailed examination of the science associated with biogenic CO_2 emissions from stationary sources to determine how to treat emissions from biomass-fired and biogenic sources. EPA charged a Scientific Advisory Board (SAB) with reviewing and commenting on (1) EPA's characterization of the science and technical issues relevant to accounting for biogenic CO_2 emissions from stationary sources; (2) EPA's framework, overall approach, and methodological choices for accounting for these emissions; and (3) options for improving upon the framework for accounting for biogenic CO_2 emissions, among other issues. [30] The SAB conducted the independent review of the agency's biogenic accounting framework and released its findings in September 2012.[31] The SAB acknowledged the "daunting task" of assessing the greenhouse gas implications of bioenergy, and the "narrow regulatory boundaries" within EPA's purview that limit the consideration of greenhouse gas flux at various points along the bioenergy pathway. The SAB identified multiple factors (e.g., time scale, spatial scale, leakage) that require further assessment by EPA and provided recommendations to revise the biogenic accounting framework. The SAB "found that quantification of most components of the framework has uncertainties, technical difficulties, data deficiencies and implementation challenges." The SAB recommended an alternative biogenic accounting factor based on feedstock category, region, land management, and prior land use. Additionally, a dissenting opinion from one of the SAB members was included in the report that would abandon the proposed framework process.

EPA also noted in the PSD and Title V Permitting Guidance for Greenhouse Gases issued in 2010 that certain types of biomass could be considered a best available control technology (BACT)—a pollution control standard mandated by the Clean Air Act—after taking into account environmental, energy, and economic considerations and state and federal policies that promote biomass for energy-independence and environmental reasons. EPA provided guidance in early 2011 on how to consider the unique GHG attributes of biomass as fuel in the BACT selection process.[32] PSD permits require that facilities apply the BACT, but BACT is determined by individual states with EPA guidance on a case-by-case basis.[33]

State perspectives on the inclusion of emissions from biomass combustion in the tailoring rule are divided.[34] Some states contend that the inclusion of biomass combustion will lead to excessive permitting requirements and fees that jeopardize renewable energy development. Other states argue that not including biomass combustion will contribute to GHG emissions over time.

Advocates of not exempting biomass combustion from the tailoring rule assert that not all biomass is carbon neutral.[35] They point out that some types of biomass, particularly biomass coming from waste streams, settle closer to the carbon-neutral and carbon-negative side of the scale. However, they assert that cutting down trees from a forest to burn in a power plant without regard to replenishing the tree stand is carbon positive. Moreover, these advocates argue, fossil fuels are still used to farm, harvest, and transport the biomass for biopower purposes, potentially negating carbon neutrality over the life cycle.

Advocates of a complete biomass combustion exemption from the tailoring rule contend that biopower plant emissions add no new carbon to the atmosphere because biopower plants would use only residuals, byproducts, and thinnings, or waste materials that would decay.[36] Furthermore, they argue that CO_2 released during biomass combustion is carbon neutral because it is re-absorbed by growing biomass. Thus, they say, measuring the emissions released during biomass combustion does not capture the entire biomass emission portfolio.

These competing interests may continue to be concerned with the designation of biomass combustion as carbon neutral because of congressional discussions and proposals to expand the biomass definition in energy legislation. Expanding the biomass definition could increase the amount of land eligible for biomass removal.[37] The biomass definition in the Energy Independence and Security Act of 2007 (EISA, P.L. 110-140) for the Renewable Fuel Standard (RFS) excludes biomass removal from federal lands, and crops from forested lands are excluded as a biofuel feedstock.[38] However, the Food, Conservation, and Energy Act of 2008 (2008 farm bill, P.L. 110-246) includes biomass from federal lands as a biofuel feedstock. The RFS addresses the carbon balance issue of liquid transportation biofuels by requiring advanced biofuels to have lower life cycle emissions relative to petroleum products. EPA was responsible for determining how the LCA would be carried out. The debate about how EPA should address the LCA, especially the land use component, was controversial.[39] While the RFS focuses on liquid transportation fuels, legislation was introduced in the 111th Congress to create a renewable electricity standard (RES).[40] If Congress debates an RES that includes biomass energy, many of the same biomass and thus carbon-neutrality concerns may emerge.

Manomet Biomass Study

The Biomass Sustainability and Carbon Policy Study by the Manomet Center for Conservation Sciences examined the use of forest biomass to generate energy for Massachusetts.[41] It studied the GHG impact of substituting one biomass feedstock type—forest biomass—for fossil fuels in the Massachusetts energy sector. More specifically, the study addressed three concerns: atmospheric GHG implications of forest biomass, forest biomass supply to support bioenergy development, and potential ecological impacts (including sustainability) associated with forest biomass harvests. Some of the study's conclusions for Massachusetts biomass include the following:[42]

- For conventional technologies, more GHGs are emitted per unit of energy produced from forest biomass than from fossil fuels.
- As a result of the excess emissions, referred to in the study as biomass "carbon debt," using forest biomass for energy initially can lead to increases in atmospheric GHG levels. However, the carbon debt is reduced as the harvested forest grows back, and at some point biomass energy yields benefits, or biomass "carbon dividends," in the form of lower GHG levels than fossil fuels.
- The time needed to pay off the carbon debt and accrue carbon dividends (or GHG benefits) can vary by decades. This timing is a function of the characteristics of the bioenergy combustion technology, the fossil fuel technology it replaces, and the biophysical and forest management characteristics of the forests from which the biomass is harvested.
- GHG benefits can be realized relatively quickly when only waste wood—for example, logging debris that would otherwise have been left in the forest—is burned for energy. But when using biomass from harvests of live trees that would otherwise have continued to grow,

GHG benefits may only be realized after many decades. Due to their efficiency, small-scale thermal and combined heat and power technologies generally yield GHG benefits sooner than utility-scale biomass electricity plants. However, where biomass replaces natural gas, the time required to achieve GHG benefits can be very long—upwards of 100 years in some cases.

- Sustainability issues are associated with forest biomass supply at the stand level (e.g., maintenance of soil productivity and biodiversity) and the landscape level (e.g., aesthetics, recreation, tourism, and long-term health of the wood-products sector).

The study's findings received considerable press attention, but a smaller spotlight was placed on the study's assumptions. Critics could argue that at least one prominent assumption renders the study irrelevant to current and future nationwide forest energy operations: the biomass feedstock type. The analysis was limited to one type of biomass—actively managed natural forests. The study found that woody biomass generated solely from logging debris (tops and branches) will contribute minimally to commercial-scale biomass facilities in Massachusetts, which "implies that the only way to meet demand would be to increase annual forest harvest." Critics might counter that throughout the United States, a range of biomass feedstocks are available for energy generation, many of which will have different greenhouse gas emission implications than actively managed natural forests.[43] The study stated "Our results for biomass from natural forests likely understate the benefits of biomass energy development relative to facilities that would rely primarily on these other wood feedstocks [tree care and landscaping sources, biomass from land clearing, and construction and demolition materials]." In some cases, if biomass feedstock types other than natural forest are used, such as woody matter that will decay, the GHG impact may be alleviated by removing material that would eventually decompose and release more potent greenhouse gases (e.g., methane). There are differing opinions on whether a single feedstock type will be the dominant feedstock for biopower, and which feedstock type (e.g., trees from natural forest, logging debris, dedicated energy crops) this could be. Assessment of national GHG implications of the various feedstock options is an area for further study.

Another study assumption that may not apply to the nationwide debate is the time required to pay off the carbon debt. GHG emission levels vary depending on the biopower pathway stage.[44] The Manomet Biomass Study compared GHG emissions per unit of energy produced by each technology for biomass, coal, oil, and natural gas. The comparisons include both direct combustion emissions as well as indirect emissions related to feedstock production, processing, and transportation. The study estimated that at the fuel combustion stage of utility-scale electricity generation, biomass emits significantly more GHGs relative to coal.[45] However, over time, biomass combustion for electricity reaches an equal cumulative GHG flux with that of fossil fuel technology (*Table 1*). Cumulative GHG emissions from biomass for electricity then level off at a rate significantly less than that of fossil fuels (*Figure 3*). Thus biomass combustion for electricity may not be the most effective method to immediately reduce GHG emissions in the short term, but may be a viable technique for GHG emission reduction over a long period. Lastly, other aspects of the study outside the scope of this report (e.g., conversion technology, biomass feedstock source, fossil fuel to be displaced) may require further examination if the findings are to be applied more broadly.

Table 1. Years to Achieve Equal Cumulative Flux with Fossil Fuels

Harvest Scenario		Fossil Fuel Technology		
	Oil (#6), Thermal	Coal, Electric	Gas, Thermal	Gas, Electric
Mixed Wood	15-30	45-75	60-90	>90
Logging Residues Only	<5	10	10	30

Source: T. Walker, "USDA Bioelectricity and GHG Workshop," Oral Presentation – "Manomet & Biomass: Moving Beyond the Soundbite," Washington, DC, November 15, 2010.

[a]. Cumulative flux is a tally of the emission and sequestration that occurs during the electricity pathway.

IS BIOPOWER CARBON NEUTRAL? IT DEPENDS

Carbon neutrality for biopower is most accurately calculated based on the carbon flux (GHG emission or sequestration) of several parameters over a specified time period. These parameters include at least the following: (1) the feedstock type; (2) the management and procurement of the energy source (in the case of biomass, how the feedstock is managed and harvested); (3) the feedstock transportation method; (4) the energy generation technology; and (5) the timeframe to replenish the feedstock. Carbon flux attributed to the management and procurement of biomass feedstock deviates according to the type or mixture of feedstocks used. For instance, agricultural biomass entails a different nutrient management plan than woody biomass. GHG emissions may be higher for agricultural biomass due to fertilizer treatments (e.g., nitrous oxide emissions from biofuel-dedicated crops).[46] Carbon flux will also vary given how the biomass feedstock is harvested.

For example, removal of woody biomass (e.g., thinnings) in large quantities may reduce carbon, and some methane, emissions on a CO_2 equivalent basis that would have been released if the woody biomass had been left in the forest to decompose. Biomass feedstock transport emits differing amounts of GHGs depending on how far the feedstock must be transported and if fossil fuels are used.[47] The carbon flux of the biopower generation technology will depend on the type of technology and if the emissions are captured and sequestered.[48]

Last, the time frame (e.g., 40 years, 100 years) assigned for biomass feedstock replenishment will determine CO_2 sequestration rates to balance out the GHGs emitted during biomass combustion, particularly for woody biomass, as growth periods (rotation ages for the trees) are often measured in decades.[49]

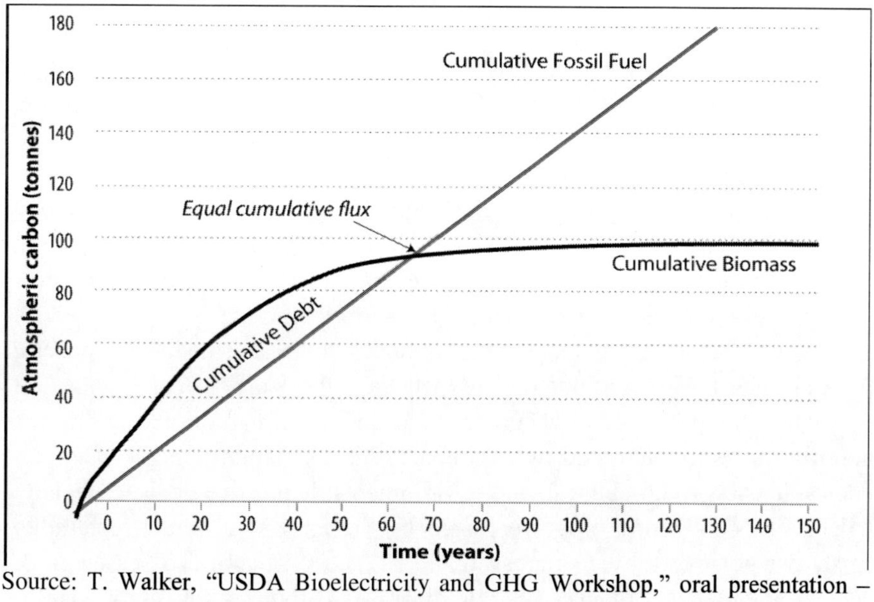

Source: T. Walker, "USDA Bioelectricity and GHG Workshop," oral presentation – "Manomet & Biomass: Moving Beyond the Soundbite," Washington, DC, November 15, 2010. Adapted by CRS.

Figure 3. Cumulative GHG Flux for Biomass and Fossil Fuel.

It could be argued that only an LCA for each biopower operation can truly determine whether biopower generation is carbon neutral. Such an LCA would measure carbon flux for each phase of the biopower pathway and incorporate biomass feedstock replenishment. A standard approach to perform a biopower LCA could ensure uniformity in GHG accounting across the biopower sector. However, multiple LCAs can be expensive and time-consuming to complete.

Biopower's carbon neutrality is a contentious aspect of the bioenergy debate. One reason the topic is so controversial is the concern about unsustainable harvests of biomass feedstocks. Some contend that if biopower proceeds with no carbon balance restrictions, large amounts of woody biomass, for example, will be removed for energy production. Another reason for controversy is concern about the air quality of areas surrounding biopower plants. These two concerns— sustainability and air quality—can be, and in some cases already are, addressed through other avenues (e.g., sustainability requirements, air quality regulations) at the federal and state level.

Legislative Implications

Biopower thus far has been included in legislation to provide energy independence and security, and under the premise that it is carbon neutral. Recent developments may prompt Congress to further analyze this premise. Carbon neutrality depends in part on the feedstock type, the technology, and the time frame for feedstock replenishment. Biopower can be created using multiple biomass feedstocks and technologies. Each feedstock and technology has its own environmental footprint. The time frame to analyze carbon neutrality is relevant because it incorporates feedstock replenishment, and thus CO_2 removal rates, and considers technology developments.

To the extent carbon neutrality continues to be a legislative concern, Congress faces a determination of whether the current carbon neutral assumption for biopower is adequate, or if additional carbon accounting for biopower is warranted and what impact this accounting might have on renewable energy, agricultural, and environmental legislative goals, especially if decisions concerning biopower made by the executive branch contradict legislative goals set by Congress (e.g., RES). A scientific assessment of whether a biopower plant is carbon positive, carbon negative, or carbon neutral may require an analysis, such as an LCA. There are time and economic constraints in conducting an LCA for each biopower operation.

Congress could decide to use existing legislative authorities to address carbon accounting for biopower. Federal environmental regulatory controls exist for the three chief environmental concerns associated with a biopower plant—air quality, land use, and water discharges. GHG emissions may be accounted for with federal regulations regarding air quality. In addition, a biopower plant also has to meet state regulatory standards, which in some cases may be stricter than the federal regulatory controls.

A full carbon accounting for biopower may result in slowing the achievement of multiple renewable energy, agricultural and environmental goals. Alternatively, the carbon neutrality debate for biopower may lead to requests for carbon accounting of some or all energy ventures— renewable and conventional. Lastly, an ill-defined carbon accounting assessment for biopower may limit public and private investment, feedstock production, and more. Scientists, investors, biomass producers, and others may hesitate to expend time and money on expanding biopower efforts if they are not provided certainty about the future contribution of biopower to U.S. energy and environmental goals.

If Congress chooses to address energy security and GHG emission increases, some have argued that these goals could be met through the creation of a national RES or a clean electricity standard (CES). The mandate of a potential national RES or CES may require substantial quantities of baseload power which some see as achieved by using biopower. If biopower is a part of an RES or CES, and if Congress continues its ongoing debate about some farm bill energy programs, the carbon neutrality designation of biopower may need to be reconsidered in response to environmental and sustainability concerns.

End Notes

[1] For more information, see CRS Report R41440, Biomass Feedstocks for Biopower: Background and Selected Issues, by Kelsi Bracmort.

[2] Biopower is a baseload power source offering "firm" power without the need for power storage. Combustion—the burning of biomass in a power plant—is the dominant technology used to produce biopower.

[3] The life cycle of a bioenergy pathway includes all stages of fuel and feedstock production and distribution, from feedstock generation or extraction through distribution, delivery, and use of the finished fuel by the ultimate consumer. The mass values for all greenhouse gases are adjusted to account for their relative global warming potential.

[4] U.S. Environmental Protection Agency, "Deferral for CO2 Emissions From Bioenergy and Other Biogenic Sources Under the Prevention of Significant Deterioration (PSD) and Title

V Programs: Final Rule," July 1, 2011. http://www.epa.gov/NSR/documents/ Biogenic_Deferral_pre-pub.pdf.

[5] Congress addressed carbon accounting for another major bioenergy pathway—liquid transportation biofuels—with a life cycle emission analysis (a requirement within the Renewable Fuel Standard). For more information, see CRS Report R40460, Calculation of Lifecycle Greenhouse Gas Emissions for the Renewable Fuel Standard (RFS), by Brent D. Yacobucci and Kelsi Bracmort.

[6] For more information, see CRS Report RL34059, The Carbon Cycle: Implications for Climate Change and Congress, by Peter Folger. Carbon is an elemental building block of molecules that make up all organisms on Earth. Carbon cycling is the process by which living things absorb carbon from the atmosphere, carbonate rocks and ocean deposits, dead organic matter in the soil, or food and return it to the atmosphere or soil by respiration, combustion, or decay.

[7] R. Miner, "Biomass 'Neutrality' in the Context of Forest-based Fuels and Products," USDA Bioelectricty and GHG Workshop, Washington, DC, November 15, 2010. Some of the definitions are not mutually exclusive.

[8] GHGs from bioenergy production can be accounted for using a life-cycle assessment (LCA). The LCA is further discussed in "Greenhouse Gas Accounting for Biopower Production," below.

[9] National Renewable Energy Laboratory, Energy Analysis, October 2010, http://www.nrel.gov/ analysis/ tech_bio_analysis.html.

[10] U.S. Environmental Projection Agency, Life Cycle Assessment: Principles and Practice, EPA/600/R-06/060, Cincinnati, OH, May 2006, http://www.epa.gov/nrmrl/lcaccess/pdfs/ 600r06060.pdf.

[11] Most LCAs for bioenergy have focused on GHG emissions from biomass used for liquid transportation fuels and its impact on climate.

[12] A cradle to grave time frame generally includes all phases from feedstock production to energy end use. A cradle to gate time frame generally includes a fraction of the complete biopower pathway which may include feedstock production, feedstock cultivation, feedstock transport, and electricity generation.

[13] The analysis did not consider land use change. NREL, Biopower Greenhouse Gas Emissions inthe LCA Literature, October 5, 2011, http://lcacenter.org/lcaxi/final/446.pdf.

[14] Q. Zhang, K. R. Goldstein, and J. R. Mihelcic, "A Review of Life Cycle Assessment Renewable Energy Derived from Forest Resources," in Renewable Energy from Forest Resources in the United States, ed. Barry D. Solomon, Calerie A. Luzadis (New York: Routledge, 2009). Information regarding the feedstocks, conversion processes, end products, system boundaries, allocation methods, and impact metrics for each LCA is available in Table 8.1.

[15] D. Peterson and S. Haase, Market Assessment of Biomass Gasification and Combustion Technology for Small- and Medium-Scale Applications, U.S. Dep. of Energy National Renewable Energy Laboratory, NREL/TP-7A2-46190, July 2009, http://www.nrel.gov/ docs/fy09osti/46190.pdf.

[16] M. K. Mann and P. L. Spath, "A Life Cycle Assessment of Biomass Cofiring in a Coal-Fired Power Plant," Clean Technologies and Environmental Policy, vol. 3, no. 2 (2001).

[17] Some coal gasification plants are under consideration to be constructed over the next few years.

[18] M. K. Mann and P. L. Spath, Life Cycle Assessment of Biomass Gasification Combined-Cycle System, Department of Energy National Renewable Energy Laboratory, December 1997, http://www.nrel.gov/docs/legosti/fy98/23076.pdf.

[19] There are other studies that examine carbon classification and feedstock sustainability for biopower. The Manomet Study is profiled in this section because it has received the most media attention thus far.

[20] May 17, 2010, http://www.saveamericasforests.org/Forests%20-%20Incinerators%20-%20Biomass/Documents/ Forests-Carbon%20Sequestration/90scientistsletter.pdf.

[21] Society of American Foresters, Scientist/Academic Letter on Biomass Energy, July 20, 2010, http://www.safnet.org/ documents/biomass_science_letter_SENATE7-20-10.pdf.

[22] Environmental Protection Agency, "Prevention of Significant Deterioration and Title V Greenhouse Gas Tailoring Rule," 75 Federal Register 31514, June 3, 2010.

[23] Carbon dioxide-equivalent (CO2 eq. or CO2e) is the principal unit of measurement across the greenhouse gases. Different gases are converted into CO2e by multiplying a quantity of the gas by its Global Warming Potential (GWP). GWP is an estimate of the relative greenhouse effectiveness over time relative to CO2 which has a GWP value of 1.

[24] For more information on the history of the tailoring rule, see CRS Report R41103, Federal Agency Actions Following the Supreme Court's Climate Change Decision in Massachusetts v. EPA: A Chronology, by Robert Meltz.

[25] Environmental Protection Agency, "Prevention of Significant Deterioration and Title V Greenhouse Gas Tailoring Rule," 75 Federal Register 31590-31591, June 3, 2010. For more information on the final rule, see CRS Report R41212, EPA Regulation of Greenhouse Gases: Congressional Responses and Options, by James E. McCarthy.

[26] U.S. EPA, "Prevention of Significant Deterioration," 75 Fed. Reg. 31590, June 3, 2010.

[27] U.S. EPA, Call for Information on Greenhouse Gas Emissions Associated with Bioenergy and Other Biogenic Sources, July 9, 2010, http://www.epa.gov/climatechange/ emissions/ biogenic_emissions.html.

[28] D. Kruger, EPA, "Panelist, Carbon Neutrality," oral presentation at Ag Energy Symposium, Washington, DC, November 4, 2010. EPA's response to the comments received is provided in U.S. Environmental Protection Agency, Deferral for CO2 Emissions from Bioenergy and Other Biogenic Sources under the Prevention of Significant Deterioration (PSD) and Title V Programs Summary of Public Comments and Responses, Washington, DC, June 28, 2011, http://www.epa.gov/NSR/documents/RTC_6-30_final_comb.pdf.

[29] U.S. Environmental Protection Agency, "Deferral for CO2 Emissions From Bioenergy and Other Biogenic Sources Under the Prevention of Significant Deterioration (PSD) and Title V Programs: Final Rule," July 1, 2011, http://www.epa.gov/NSR/documents/ Biogenic_ Deferral_pre-pub.pdf. Biogenic includes facilities that emit CO2 from sources originating via biological processes, such as landfills.

[30] The agency's charge for the SAB, review documents including the accounting framework, and meeting materials are available at http://yosemite.epa.gov/sab/sabproduct.nsf/0/ 2f9b572c712ac52e8525783100704886!OpenDocument& TableRow=2.2#2.

[31] U.S. Environmental Protection Agency, SAB Review of EPA's Accounting Framework for Biogenic CO2 Emissions from Stationary Sources, EPA-SAB-12-011, September 28, 2012.

[32] U.S. Environmental Protection Agency, Guidance for Determining Best Available Control Technology for Reducing Carbon Dioxide Emissions from Bioenergy Production, Washington, DC, March 2011, http://www.epa.gov/NSR/ ghgdocs/bioenergyguidance.pdf.

[33] Environmental Protection Agency, PSD and Title V Permitting Guidance For Greenhouse Gases, November 2010, http://www.epa.gov/nsr/ghgdocs/epa-hq-oar-2010-0841-0001.pdf.

[34] Energy Washington, States Split on Whether Biomass Should Be Exempt from GHG Permits, September 22, 2010.

[35] N. Greene, Scientists to Congress & Obama: Count the Carbon in Biomass, Natural Resources Defense Council, May 24, 2010, http://switchboard.nrdc.org/blogs/ngreene/ scientists_to_ congress_obama_c.html.

[36] Personal communication with B. Cleaves, CEO, Biomass Power Association, October 1, 2010.

[37] For more information, see CRS Report R40529, Biomass: Comparison of Definitions in Legislation Through the 112th Congress, by Kelsi Bracmort.

[38] The Renewable Fuel Standard (RFS) is a provision established by the Energy Policy Act of 2005 requiring gasoline to contain a minimum amount of fuel produced from renewable biomass. For more information on the RFS, see CRS Report R40155, Renewable Fuel Standard (RFS): Overview and Issues, by Randy Schnepf and Brent D. Yacobucci.

[39] For more information, see CRS Report R40460, Calculation of Lifecycle Greenhouse Gas Emissions for the Renewable Fuel Standard (RFS), by Brent D. Yacobucci and Kelsi Bracmort.

[40] For more information on the renewable electricity standard debate, see CRS Report R40565, Biomass Resources: The Southeastern United States and the Renewable Electricity Standard Debate, by Richard J. Campbell.

[41] The Manomet Center for Conservation Sciences is a non-profit research organization that works to integrate society's social, economic, and environmental values to create sustainable systems for present and future generations.

[42] Personal communication with T. Walker, November 30, 2010.

[43] For more information on the different kinds of biomass available for power generation, see CRS Report R41440, Biomass Feedstocks for Biopower: Background and Selected Issues, by Kelsi Bracmort.

[44] The various biopower pathway stages are explained in the "Greenhouse Gas Accounting for Biopower Production" section of this report.

[45] T. Walker, P. Cardellichio, and A. Colnes, et al., Biomass Sustainability and Carbon Policy Study, Manomet Center for Conservation Sciences, NCI-2010-03, June 2010, p. 103, http://www.manomet.org/sites/manomet.org/files/ Manomet_Biomass_Report_Full_LoRez.pdf.

[46] When biomass used for biopower production is agricultural waste, some GHG emissions may be attributed to crop cultivation whereby the crop is used for other feed, fiber and fuel purposes.

[47] Some make the case that fuels (e.g., ethanol) other than fossil fuels (e.g., diesel or gasoline) may be used to transport the feedstock.

[48] There are no commercial carbon capture and sequestration(CCS) projects in operation at the moment. Therefore, CCS is not likely to impact carbon flux at the biopower generation stage in the near-term.

[49] For more information on carbon sequestration in trees, see CRS Report R40562, U.S. Tree Planting for Carbon Sequestration, by Ross W. Gorte.

INDEX

A

B

C

D

E

F

G

H

I

N

O

P

Q

R

S

T

U

V

W

Y